To my father, He Meng-Qing

Tian-Xiao He

Dimensionality Reducing Expansion of Multivariate Integration

Birkhäuser
Boston • Basel • Berlin

Tian-Xiao He
Department of Mathematics & Computer Science
Illinois Wesleyan University
Bloomington, IL 61702-2240
U.S.A.

Library of Congress Cataloging-in-Publication Data

He, Tian-Xiao, 1952-
Dimensionality reducing expansion of multivariate integration / Tian-Xiao He.
p. cm.
Includes bibliographical references and index.
ISBN 0-8176-4170-X (acid-free paper) – ISBN 3-7643-4170-X (acid-free paper)
1. Numerical integration. 2. Gaussian quadrature formulas. 3. Green's functions. I.
Title.
QA299.3.H4 2001
515'.624–dc21

2001025166

AMS Subject Classifications: 65D30, 65D32, 34B27, 26B20, 41A50

Printed on acid-free paper.

Birkhäuser 

ISBN 0-8176-4170-X SPIN 10754122
ISBN 3-7643-4170-X

Reformatted from author's files in LATEX2ε by TEXniques, Inc., Cambridge, MA
Printed and bound by Hamilton Printing, Rensselaer, NY
Printed in the United States of America

9 8 7 6 5 4 3 2 1

Contents

Preface

This book is a self-contained monograph on numerical multivariate integration. Although it has been a well-established subject for some time, numerical multivariate integration has never been applied as widely as it is now. We can find its applications almost everywhere in mathematics, economics, and the natural sciences. A good example is the collateralized mortgage obligation (CMO), which is a bundle of loans that generates cash flows from interest payments and the repayment of principal. It is clear that fluctuations in interest rates alter the cash flows. Let us consider a CMO consisting of 15-year mortgages with monthly payments. There are 180 cash flows in all, depending on 180 interest rates, which are basically 180 variables. The goal is to compute the present expected value of the CMO, averaged over all possible fluctuations of the 180 interest rates. Parameterized by the probabilities with which fluctuations occur, this calculation can be formulated as a multivariate integral over the 180-dimensional unit cube ([10]).

This book discusses a technique for numerical integration by using *dimensionality reducing expansions* (*DRE*) to reduce a higher dimensional integral to lower dimensional integrals with or without a remainder. Obviously, a DRE can be used to reduce the computational load of many very high dimensional numerical integrations. For instance, the multivariate integral in the CMO problem can be reduced to integrals of any lower dimension by successive applications of certain DREs; it can even be directly reduced to a one-dimensional integral by a single application of the measure theory. In most cases, the computation needed for the reductions and the final integration is miniscule compared to that of the original integration.

Most DREs are based on *Green's Theorem* in the real or complex field. In 1963, using the theorem, Hsu [37] devised a way to construct a DRE with algebraic precision (degree of accuracy) for multivariate integrations. From 1978 to 1986, building on [37], Hsu, Zhou, and the author (see [39], [42], [43], and [44]) developed a general process for constructing a DRE with algebraic precision and estimating its remainder. In 1972, with the aid of Green's Theorem and the *Schwarz function*, P.J. Davis [15] gave an exact DRE, or a DRE without a remainder, for a double integral over a complex field. In 1979, also utilizing Green's Theorem, Kratz [51] constructed an exact DRE for functions that satisfy a specific type of partial differential equations. Lastly, to complete the introduction, we mention Burrows' DRE for measurable functions, developed in the 1980s. As noted above, his DRE can reduce a multivariate integration to a one-dimensional integral in a single step. Some important and common applications of DRE include the construction of boundary type quadrature formulas and asymptotic formulas for oscillatory integrals. A *boundary type quadrature formula (BTQF)* is an approximate integration formula with all of its evaluation points lying on the boundary of the integration domain. Such a formula is particularly useful for cases where the values of the integrand function and its derivatives inside the domain are not given or are not easily determined. Boundary quadrature formulas are not really new. Indeed, from the viewpoint of numerical analysis, the classic *Euler–Maclaurin summation formula* and the *Hermite two-end multiple nodes quadrature formula* may be regarded as one-dimensional BTQFs since they use only the values of the integrand function and their derivatives at the limits of integration. The earliest example of a BTQF with some algebraic precision for multivariate integration is probably the formula of algebraic precision degree 5 for a triple integral over a cube given by Sadowsky [60] in 1940. He used 42 points on the surface of a cube to construct the quadrature, which has been modified by the author to one with 32 points, the fewest possible boundary points (see [27] and [28]). Some 20 years later, Levin [52] and [53], Federenko [21], and Ionescu [47] individually investigated certain optimal BTQFs for double integration over squares using partial derivatives at some boundary points of the region. Despite these advances, however, both the general principle and the general technique of construction remained lacking for many years.

Since 1978, Hsu, Zhou, Wang, Yang, and the author have developed several general methods for constructing BTQFs using the basic principles of multivariate integration DRE ([27], [28], [30], [32], [39], [40], [41], [42], [43], and [44]). In this book, we will describe all of these results. We will also look at many recent developments in this field, such as using BTQFs to develop a boundary element scheme, determining the fewest possible evaluation points of a BTQF, and constructing BTQFs associated with wavelet functions.

Another application of DREs is in the evaluation of *oscillatory integrals*. In 1947, the physicist A. Maréchal developed a mechanical quadrature to study the distribution of light. His method for constructing the quadrature is based on the idea of expressing a double integral by the limit of line integrals. Inspired by this, we develop a type of DRE such that a $2k$-dimensional integral can be expressed as limits of k-dimensional integrals. These DREs can help us to numerically evaluate oscillatory integrals, to which the conventional techniques involving polynomial approximations are difficult to apply. Denote by $\langle x \rangle$ the fractional part of the nonnegative real number x, and let $N_1, \ldots, N_k$ be infinitely large parameters. Then the main oscillatory integrals we are going to discuss in this book are integrals with the form $\int_V f(X; \langle NX \rangle dV$, where $X = (x_1, \ldots, x_k)$, $\langle NX \rangle = (\langle N_1 x_1 \rangle, \ldots, \langle N_k x_k \rangle)$, and dV is the volume measure in V. The book also contains results on other oscillatory integrals and recent results such as Bradie, Coifman, and Grossmann's work in [2].

This book is not only a reference for numerical analysis researchers and scientists but can also be used as a textbook for graduate and advanced undergraduate students. It is arranged as follows. Chapters 1 and 2 discuss the construction of DREs and BTQFs with various degrees of algebraic precision. In Chapters 3 and 4, we use DREs to approximate oscillatory integrals and establish their corresponding numerical quadrature formulas. Chapter 5 demonstrates how to construct DREs over a complex region by using the Schwarz function and the *Bergman kernel.* Chapter 6 concludes the book by examining how the solutions of certain differential equations can be used to construct exact DREs as well as how, conversely, some DREs can be utilized to derive a scheme of the boundary element method, used for evaluating the numerical solution of a boundary value problem of a partial differential equation. The writing of this monograph was greatly influenced by the pioneering work of B. Bradie, B.L. Burrows, R. Coifman, P.J. Davis, A. Ghizzetti, W. Gröbner, A. Grossmann, T. Havie, L.C. Hsu, L.J. Kratz, A. Ossicini, P. Rabinowitz, D.D. Stancu, A.H. Stroud, G.G. Walter, and Y.S. Zhou. I would like to express thanks for the assistance from Henry He, who typed a portion of the manuscript and prepared it in camera-ready form. I am also indebted to G.A. Anastassiou, a guest editor at Birkhäuser Boston, for letting me know about this opportunity. Finally, to the editorial staff at Birkhäuser, particularly Ann Kostant and Tom Grasso, I wish to express my sincere appreciation for their efficient assistance and patience.

Tian-Xiao He
February 2001

Dimensionality Reducing Expansion of Multivariate Integration

Chapter 1

Dimensionality Reducing Expansion of Multivariate Integration

In this chapter, we will discuss DRE construction using the generalized integration by parts rule, DREs with algebraic precision, and the minimum estimation of the remainders in DREs with algebraic precision. Some optimal DREs will also be given. As the starting point, we will look at *Darboux formulas*, which can be considered as one-dimensional DREs.

1.1 Darboux formulas and their special forms

The DRE of multivariate integration resulted from the idea behind the derivation of the Darboux formula, which is thus a good starting point for a discussion on the DRE.

The Darboux formula, first given in 1876, is an expression formula for an analytic function. The Taylor formula is one of its special cases.

Let $f(z)$ be normal analytic on the line connecting points a and z, and $\phi(t)$ a polynomial of degree n. The derivative of $\phi(t)$ with respect to t $(0 \leq t \leq 1)$ can be written as

$$
\begin{aligned}
& \frac{d}{dt}\sum_{v=1}^{n}(-1)^{v}(z-a)^{v}\phi^{(n-v)}(t)f^{(v)}(a+t(z-a)) \\
= \; & -(z-a)\phi^{(n)}(t)f'(a+t(z-a)) \\
& +(-1)^{n}(z-a)^{n+1}\phi(t)f^{(n+1)}(a+t(z-a)) \, .
\end{aligned}
$$

By taking the integral in terms of t from 0 to 1 and noting that $\phi^{(n)}(t) = \phi^{(n)}(0)$, we obtain

$$\begin{aligned}
&\phi^{(n)}(0)[f(z)-f(a)] \\
&= \sum_{v=1}^{n}(-1)^{v-1}(z-a)^{v}\left[\phi^{(n-v)}(1)f^{(v)}(z)-\phi^{(n-v)}f^{(v)}(a)\right] \\
&\quad +(-1)^{n}(z-a)^{n+1}\int_0^1 \phi(t)f^{n+1}(a+t(z-a))dt. \qquad (1.1.1)
\end{aligned}$$

Equation (1.1.1) is the Darboux formula. Now we will consider several of its special cases.

Example 1.1.1. Let $\phi(t) = (t-1)^n$ in equation (1.1.1). Then $\phi^{(n)}(0) = n!$, $\phi^{(n-v)}(1) = 0$, and $\phi^{(n-v)}(0) = (-1)^v n!/v!$, for $1 \le v \le n$. This is of course the Taylor formula with a Cauchy integral form remainder.

Example 1.1.2. In equation (1.1.1), changing n to $2n$, setting $\phi(t) = t^n(t-1)^n$, and taking $n \to \infty$ yields

$$f(z)-f(a)=\sum_{v=1}^{\infty}\frac{(-1)^{v-1}(z-a)^{v}}{2^{v}\cdot v!}[f^{(v)}(z)+(-1)^{v-1}f^{(v)}(a)].$$

Here the series is assumed to be convergent.

Without loss of generality, let the highest power term of $\phi(t)$ be t^n and $F(t)$ be an antiderivative of $f(t)$. $F(t)$ is assumed to be n order continuously differentiable. Then the following integral quadrature formula will be obtained by setting $a = 0$ and $z = 1$ in equation (1.1.1):

$$\begin{aligned}
\int_0^1 F(t)dt &= \sum_{v=1}^{n}\frac{(-1)^{v-1}}{n!}\left[\phi^{(n-v)}(t)F^{(v-1)}(t)\right]_{t=0}^{t=1} \\
&+ \frac{(-1)^n}{n!}\int_0^1 \phi(t)F^{(n)}(t)dt\,. \qquad (1.1.2)
\end{aligned}$$

In fact, equation (1.1.2) can easily be verified by applying integration by parts n times to its integral form remainder. One may rewrite the above formula into a more general form (a suitable transformation of variable is needed).

$$\int_a^b F(t)dt=\sum_{v=1}^{n}\frac{(-1)^{v-1}}{n!}\left[\phi^{(n-v)}(t)F^{(v-1)}(t)\right]_{t=a}^{t=b}+R_n\,, \qquad (1.1.3)$$

where remainder R_n is

$$R_n=\frac{(-1)^n}{n!}\int_a^b \phi(t)F^{(n)}(t)dt. \qquad (1.1.4)$$

Equations (1.1.3) and (1.1.4) can be understood as the integral form of the Darboux formula. However, the applications of the Darboux formula to numerical integration were given attention only after 1940 (see [57]).

The formula shown in equation (1.1.3) has two features: (1) Except for the remainder, the right-hand side of the equation consists of the values of the integrand and its derivatives at end points of the integral interval. Hence, the formula is a BTQF. (2) By choosing a suitable weight function $\phi(t)$ in the remainder (1.1.4), we can make $R_n \equiv R_n(F)$ have the smallest possible estimate in various norms.

In addition, setting $\phi(t)$ as the nth order Bernoulli polynomial in equation (1.1.3), we obtain the *Euler–Maclaurin formula.*

Also, if $n = 2m$ and $\phi(t) = (t-a)^m(t-b)^m$, then equation (1.1.3) will give the following Petr formula when the Leibniz higher order differentiation rule is applied.

$$\begin{aligned} \int_a^b F(t)dt &= \sum_{v=1}^{m} \frac{\binom{m}{v}}{\binom{2m}{v}} \frac{(b-a)^v}{v!} \Big[F^{(v-1)}(a) \\ &+ (-1)^{v-1} F^{(v-1)}(b)\Big] + R_m, \end{aligned} \tag{1.1.5}$$

where remainder R_m is given by equation (1.1.4). Using the mean value theorem of integration, we can write it as

$$R_m = \frac{(-1)^m}{2^{m+1}} [\frac{m!}{(2m)!}]^2 F^{(2m)}(\xi)(b-a)^{2m+1}, \qquad a \le \xi \le b.$$

Similarly, setting $n = m+k$ and $\phi(t) = (a-t)^k(b-t)^m$ in equation (1.1.3), we obtain the so-called *Obreschkoff formula*

$$\begin{aligned} \int_a^b F(t)dt &= \frac{m!k!}{(m+k)!} \sum_{v=1}^{m} \binom{m+k-v}{k} \frac{(b-a)^v}{v!} f^{(v)}(a) \\ &- \frac{m!k!}{(m+k)!} \sum_{v=1}^{k} (-1)^k \binom{m+k-v}{k} \frac{(b-a)^v}{v!} f^{(v)}(b) \\ &- \frac{m!k!}{(m+k)!} R_{m+k}, \end{aligned} \tag{1.1.6}$$

where

$$R_{m+k} = -\frac{1}{k!m!} \int_a^b (b-t)^m (a-t)^k f^{(m+k+1)}(t)dt. \tag{1.1.7}$$

All of the above formulas are useful in approximating definite integrals. In Section 1.6, applications of the Obreschkoff formula to rational approximation and the *Petr formula* to eigenvalue problems will be discussed.

At the end of this section, we will give one more example of the Darboux formula. In equation (1.1.3), we set $a = -1$, $b = 1$, and $\phi(t) = T_n(t) + 1/2^{n-1}$, where $T_n(t) = 2^{1-n}\cos(n\cos^{-1} t)$ is the Chebyshev polynomial of degree n. Then,

$$\phi(1) = \frac{1}{2^{n-2}}, \qquad \phi(-1) = \frac{1+(-1)^n}{2^{n-1}}.$$

Since $T_n(t)$ satisfies differential equation $(1-t^2)T_n''(t) - tT_n'(t) + n^2T_n(t) = 0$, by taking derivatives consecutively, we obtain

$$\begin{aligned} \phi^{(n-v)}(-1) &= (-1)^v \phi^{(n-v)}(1), \\ 2^{n-1}\phi^{(n-v)}(1) &= \tfrac{n^2}{1} \cdot \tfrac{n^2-1^2}{3} \cdots \tfrac{n^2-(n-v-1)^2}{2n-2v-1}, \end{aligned}$$

where $v = 1, 2, \ldots, n-1$. Therefore, when n is an even integer, we can derive the following BTQF from equation (1.1.3).

$$\begin{aligned} &\int_{-1}^{1} F(t)dt = F(1) + F(-1) \\ &+ \sum_{v=2}^{n-1} \frac{(-1)^{v-1}}{2^v \cdot v!} \frac{\begin{pmatrix} 2n-v-1 \\ v \end{pmatrix}}{\begin{pmatrix} n-1 \\ v \end{pmatrix}} \left[F^{(v-1)}(1) + (-1)^{v-1}F^{(v-1)}(-1)\right] \\ &- \frac{1}{2^{n-2} \cdot n!}\left[F^{(n-1)}(1) - F^{(n-1)}(-1)\right] \\ &+ \left(\frac{n^2-2}{n^2-1}\right) \frac{F^{(n)}(\xi)}{2^{n-2} \cdot n!}, \end{aligned} \tag{1.1.8}$$

where $-1 \le \xi \le 1$.

It is obvious that the BTQF derived from equation (1.1.3) by deleting the remainder R_n possesses algebraic precision degree $n-1$. Here, the degree of algebraic precision of a quadrature formula is the largest positive integer r such that the formula is exact for x^i, when $i = 0, 1, \ldots, r$. Hence, choosing a suitable $\phi(t)$, we can construct various BTQFs for different purposes and make the corresponding remainders the smallest possible under different norms.

In the rest of this chapter, we will extend the idea of constructing the Darboux formula to higher dimensional settings.

1.2 Generalized integration by parts rule

Suppose that we are given a class of multiple integrals of the form

$$J := \int_\Omega F(X)\,dX, \quad F(X) \in K,$$

where K is a certain class of real functions defined on a bounded domain $\Omega \subset \mathbb{R}^n$, $X \equiv (x_1, \dots, x_n)$ represents a point-vector, and $dX \equiv dx_1, \dots, dx_n$. The $n-1$ dimensional boundary surface of Ω is denoted by $\partial\Omega$. Our goal is to obtain approximate integration formulas of the form

$$J \approx \sum_{i,j} A_{ij}\Lambda_i(F)(X_{ij}), \ X_{ij} \in \partial\Omega,$$

where each $\Lambda_i(F)$ stands for a certain linear combination of F and some of its partial derivatives, X_{ij} are evaluation points (nodes) lying on the boundary of Ω, and A_{ij} are the corresponding quadrature coefficients.

One general process of construction consists of three main steps. (1) Replace J by a set of boundary surface integrals with an error term that can be reflected or well-controlled. (2) Choose a suitable auxiliary function based on the question so as to minimize or diminish the error term. (3) Choose suitable numerical integration formulas that can be used to approximate those boundary surface integrals with a certain accuracy. Note that the first step may easily be achieved by means of successive applications of Green's formula in higher dimensions.

In general, for a differential operator L of order m, if there exists a differential operator M of the same order such that

$$\int_\Omega (F \cdot L(G) - G \cdot M(F))\,dX = \int_{\partial\Omega} W\,dS, \tag{1.2.1}$$

where W is a linear combination of the products of F and G or their partial derivatives with their total order not exceeding $m-1$, then M is called the *adjoint differential operator* of L. If M is equal to L, then L is said to be a *self-adjoint differential operator*.

In this section we will discuss a differential operator L defined by

$$L = \sum_{m_1+\cdots+m_n \le N} \lambda_{m_1,\dots,m_n} D^{(m_1,\dots,m_n)}, \tag{1.2.2}$$

where $\lambda_{m_1,\dots,m_n}$ are constants and $D^{(m_1,\dots,m_n)} = \partial^{m_1+\cdots+m_n} / \partial x_1^{m_1} \dots \partial x_n^{m_n}$. In particular, $D^{(0,\dots,0,m_j,0,\dots,0)} = (\partial/\partial x_j)^{m_j}$. All m_i, $i = 1, \dots, n$, are nonnegative integers. $m = \sum_{i=1}^n m_i$ is called the order of a differential operator $D^{(m_1,\dots,m_n)}$. If N denotes the highest of the orders of operators $D^{(m_1,\dots,m_n)}$ in operator L, then N is called the *order of operator* L. Moreover, if every operator $D^{(m_1,\dots,m_n)}$ is even order (or odd ordered), then L is an *even (or odd) operator*.

From the following lemma on the *general integration by parts formula*, we know that the adjoint operator of L is $M = (-1)^m L$.

Lemma 1.2.1 *Let $\Omega \subset \mathbb{R}^n$ be an n-dimensional bounded closed domain with its boundary $S = \partial\Omega$ being a piecewise smooth surface. In particular, if $n = 2$, $\partial\Omega$ is a simple closed curve with finite length. Let $F(X), G(X) \in C^m(\Omega)$, where m is the order of the operator L. Then*

$$\int_\Omega F \cdot L(G)\, dX = \int_{\partial\Omega} W\, dS + (-1)^m \int_\Omega G \cdot L(F)\, dX, \tag{1.2.3}$$

where dS is the area element of $\partial\Omega$ and W is a linear combination of all the possible terms formed by the products of partial derivatives of F and of G with their total order not exceeding $m - 1$.

Proof. Since L is a linear combination of $D^{(m_1,\dots,m_n)}$, it is sufficient to prove equation (1.2.3) for $L = D^{(m_1,\dots,m_n)}$. Here $m = m_1 + \cdots + m_n$. The proof depends on the *Gauss–Green formula*

$$\int_\Omega \frac{\partial f(X)}{\partial x_i} dV = \int_{\partial\Omega} f(X) \frac{\partial x_i}{\partial \nu} dS, \tag{1.2.4}$$

where $\frac{\partial x_i}{\partial \nu}$ is the outer normal derivative of x_i on the surface $\partial\Omega$. In particular, if $\partial\Omega$ can be expressed as $Q(x_1, \dots, x_n) = 0$, then

$$\frac{\partial x_i}{\partial \nu} = \frac{\partial Q}{\partial x_i} / [(\frac{\partial Q}{\partial x_1})^2 + \cdots + (\frac{\partial Q}{\partial x_n})^2]^{-\frac{1}{2}}.$$

Substituting $f(X) = F(X) \cdot G(X)$ into equation (1.2.4), we have

$$\int_\Omega F \frac{\partial G}{\partial x_i} dX = \int_{\partial\Omega} F \cdot G \frac{\partial x_i}{\partial \nu} dS - \int_\Omega G \frac{\partial F}{\partial x_i} dX. \tag{1.2.5}$$

It is simply a multivariate integration by parts formula. By repeatedly applying this formula, we obtain equation (1.2.3) with

$$\begin{aligned} W = {} & \sum_{i=1}^{n} \sum_{k=0}^{m_i - 1} (-1)^{m_1 + \cdots + m_{i-1} + k} D^{(m_1,\dots,m_{i-1},k,0,\dots,0)} F(X) \times \\ & D^{(0,\dots,0,m_i-k-1,m_{i+1},\dots,m_n)} G(X) \frac{\partial x_i}{\partial \nu}. \end{aligned} \tag{1.2.6}$$

Here, for $i = 1$ and $i = n$, we take, respectively, $D^{(m_1,\dots,m_{i-1},k,0,\dots,0)} = D^{(k,0,\dots,0)}$ and $D^{(0,\dots,0,m_i-k-1,m_{i+1},\dots,m_n)} = D^{(0,\dots,0,m_n-k-1)}$ in (1.2.6). Obviously, the order of differentiation of each term of W is $m - 1$. □

Since there exists a certain symmetry in equation (1.2.3) (i.e., F and G can be switched with each other), the general integration by parts formula (1.2.3) is also called the *symmetric rule*.

Example 1.2.1. Let $n = 3$, $L = D^{(m,m,m)} = \partial^{3m}/\partial x^m \partial y^m \partial z^m$, $F(x, y, z)$, $G(x, y, z) \in C^{3m}(\Omega)$, $\Omega \subset \mathbb{R}^3$. This is equivalent to setting $n = 3$, $M = 3m$, and $m_1 = m_2 = m_3 = m$ in equation (1.2.3). Therefore,

$$\begin{aligned}
&\int_\Omega F \cdot D^{(m,m,m)} G dxdydz - (-1)^{3m} \int_\Omega G \cdot D^{(m,m,m)} F dxdydz \\
= &\sum_{k=0}^{m-1} (-1)^k \int_{\partial\Omega} D^{(k,0,0)} F \cdot D^{(m-k-1,m,m)} G dydz \\
&+ \sum_{k=0}^{m-1} (-1)^{m+k} \int_{\partial\Omega} D^{(m,k,0)} F \cdot D^{(0,m-k-1,m)} G dzdx \\
&+ \sum_{k=0}^{m-1} (-1)^{2m+k} \int_{\partial\Omega} D^{(m,m,k)} F \cdot D^{(0,0,m-k-1)} G dxdy. \qquad (1.2.7)
\end{aligned}$$

If G is chosen as a polynomial of degree $3m$, with the coefficient of the term $x^m y^m z^m$ being 1, then we obtain the homogeneous DRE

$$\begin{aligned}
&\int_\Omega F dxdydz \\
= &\sum_{k=0}^{m-1} \frac{(-1)^k}{(m!)^3} \int_{\partial\Omega} D^{(k,0,0)} F \cdot D^{(m-k-1,m,m)} G dydz \\
+ &\sum_{k=0}^{m-1} \frac{(-1)^{m+k}}{(m!)^3} \int_{\partial\Omega} D^{(m,k,0)} F \cdot D^{(0,m-k-1,m)} G dzdx \\
+ &\sum_{k=0}^{m-1} \frac{(-1)^{2m+k}}{(m!)^3} \int_{\partial\Omega} D^{(m,m,k)} F \cdot D^{(0,0,m-k-1)} G dxdy \\
+ &\rho_m, \qquad (1.2.8)
\end{aligned}$$

where

$$\rho_m = \frac{(-1)^{3m}}{(m!)^3} \int_\Omega G \cdot D^{(m,m,m)} F dxdydz.$$

DRE (1.2.8) is said to be homogeneous or exact if its remainder ρ_m is zero. Obviously, DRE (1.2.8) is homogeneous when $F(x, y, z)$ is a polynomial of the form $p(x, y, z) = ax^m(yz)^{m-1} + by^m(xz)^{m-1} + cz^m(xy)^{m-1} +$ a polynomial of lower total degree.

We might apply the multivariate integration by parts formula (1.2.5) in terms of x, y, z, and repeat this process. Then we obtain a different DRE. For example, consider the case shown in Example 1; i.e., $n = 3$, $L = D^{(m,m,m)} =$

$\partial^{3m}/\partial x^m \partial y^m \partial z^m$, $F(x,y,z), G(x,y,z) \in C^{3m}(\Omega)$, $\Omega \subset \mathbb{R}_3$, and G is also a polynomial of degree $3m$, with the coefficient of the term $x^m y^m z^m$ being 1. We thus have a new DRE as follows:

$$\begin{aligned}
&\int_\Omega F dxdydz \\
= &\frac{1}{(m!)^3} \sum_{k=0}^{m-1} (-1)^{3k} \int_{\partial\Omega} [D^{(k,k,k)} F \cdot D^{(m-k-1,m-k-1,m-k-1)} G''_{yz} dydz \\
- &D^{(k,k,k)} F'_x \cdot D^{(m-k-1,m-k-1,m-k-1)} G'_z dzdx \\
+ &D^{(k,k,k)} F''_{xy} \cdot D^{(m-k-1,m-k-1,m-k-1)} G dxdy] + \rho_m,
\end{aligned} \tag{1.2.9}$$

where

$$\rho_m = \frac{(-1)^{3m}}{(m!)^3} \int_\Omega G \cdot D^{(m,m,m)} F dxdydz.$$

Comparing equations (1.2.8) and (1.2.9), we find that they are essentially the same. However, the first formula is more flexible in application. By using it, we can find DREs with various degrees of algebraic precision. More details can be found in the next section.

1.3 DREs with algebraic precision

In this section, we will give DREs with algebraic precision. Let $\Omega \equiv V_n$ be a bounded and closed region in $\mathbb{R}^n$. Suppose that the boundary of V_n, S_{n-1}, can be described by a system of parametric equations. In particular, points $(x_1, \ldots, x_n)$ on S_{n-1} satisfy the equation

$$\Phi(x_1, \ldots, x_n) = 0, \tag{1.3.1}$$

where Φ has continuous partial derivatives. In addition, $\Phi(x_1, \ldots, x_n) \le 0$ for all points in V_n.

Assume $F(X) = F(x_1, \ldots, x_n)$ and $G(X) = G(x_1, \ldots, x_n)$ are continuously differentiable on V_n. It follows from formula (1.2.5) that

$$\int_{V_n} F \frac{\partial G}{\partial x_n} dV_n = \int_{S_{n-1}} F \cdot G \frac{\partial x_n}{\partial \nu} dS - \int_{V_n} G \frac{\partial F}{\partial x_n} dV_n, \tag{1.3.2}$$

where $dV_n = dx_1 \cdots dx_n$. If we further assume that functions $F(X)$ and $G(X)$ have continuous mth order partial derivatives with respect to x_n and $G(X)$ is of the form

$$G(X) = x_n^m + \sum_{i=0}^{m-1} Q_i(x_1, \ldots, x_{n-1}) x_n^i, \quad Q_i \in C^m, \tag{1.3.3}$$

then repeated applications of equation (1.3.2) give us the following formula.

$$\int_{V_n} F(X)dV = \sum_{k=0}^{m-1} \frac{(-1)^k}{m!} \int_{S_{n-1}} L_k(F,G)dS + \rho_m, \tag{1.3.4}$$

where

$$L_k(F,G) \equiv \left(\frac{\partial^k F}{\partial x_n^k}\right)\left(\frac{\partial^{m-k-1} G}{\partial x_n^{m-k-1}}\right)\left(\frac{\partial x_n}{\partial \nu}\right)$$

and the remainder ρ_m can be expressed as

$$\rho_m = \frac{(-1)^m}{m!} \int_{V_n} \left(\frac{\partial^m F}{\partial x_n^m}\right) G(X)dV. \tag{1.3.5}$$

If ρ_m is deleted, equation (1.3.4) gives the approximate formula

$$\int_{V_n} F(X)dV \approx \frac{1}{m!} \sum_{k=0}^{m-1} (-1)^k \int_{S_{n-1}} \frac{\partial^k F}{\partial x_n^k} \frac{\partial^{m-k-1} G}{\partial x_n^{m-k-1}} \frac{\partial x_n}{\partial \nu} dS. \tag{1.3.6}$$

Formula (1.3.6) is a DRE with m terms. Remainder expression (1.3.5) shows that the degree of algebraic precision of the above formula is $m-1$.

If the boundary surface, S_{n-1}, of V_n is defined by parametric equations

$$x_i = x_i(t_1, \ldots, t_{n-1}), \qquad i = 1, \ldots, n,$$

then surface integrals on the right-hand side of equation (1.3.4) can be written as $(n-1)$-dimensional multivariate integrals. Furthermore, if V_n is defined by

$$\Phi_1(x_1, \ldots, x_{n-1}) \le x_n \le \Phi_2(x_1, \ldots, x_{n-1})$$

where

$$(x_1, \ldots, x_{n-1}) \in V_{n-1}$$

and Φ_1 and Φ_2 are continuous functions, then the basic formula (1.3.4) can be reduced to

$$\int_{V_n} F(X)dV_n$$
$$= \sum_{k=0}^{m-1} \frac{(-1)^k}{m!} \int_{V_{n-1}} \left[\frac{\partial^k F}{\partial x_n^k} \frac{\partial^{m-k-1} G}{\partial x_n^{m-k-1}}\right]_{x_n=\Phi_1}^{x_n=\Phi_2} dV_{n-1} + \rho_m, \tag{1.3.7}$$

where $dV_n = dx_1, \ldots, dx_n$ and

$$[g(X)]_{x_n=\Phi_1}^{x_n=\Phi_2} = g(x_1, \ldots, x_{n-1}, \Phi_2) - g(x_1, \ldots, x_{n-1}, \Phi_1).$$

Equation (1.3.7) shows that an n-dimensional multivariate integral over bounded region V_n can be expressed in terms of finitely many $(n-1)$-dimensional multivariate integrals over V_{n-1} with the remainder ρ_m shown as expression (1.3.5).

Obviously, the Darboux formula (1.1.3) is a special case of formula (1.3.7) for $n = 1$.

Denote $\|f\|_C = \max_{X \in V_n} |f(X)|$, $\|f\|_{L_1} = \int_{V_n} |f(X)| dV$, and $\|f\|_{L_2} = \left(\int_{V_n} |f(X)|^2 dV\right)^{\frac{1}{2}}$. From expression (1.3.5), we have the estimates

$$|\rho_m| \le \frac{1}{m!} \left\| \frac{\partial^m F}{\partial x_n^m} \right\|_C \|G(X)\|_{L_1}, \tag{1.3.8}$$

$$|\rho_m| \le \frac{1}{m!} \left\| \frac{\partial^m F}{\partial x_n^m} \right\|_{L_2} \|G(X)\|_{L_2}, \tag{1.3.9}$$

and

$$|\rho_m| \le \frac{1}{m!} \left\| \frac{\partial^m F}{\partial x_n^m} \right\|_{L_1} \|G(X)\|_C. \tag{1.3.10}$$

Although there exist many functions of form (1.3.3) that can potentially serve as the auxiliary function $G(X)$ in DREs, we should concentrate on those that satisfy the following three principles. First the remainder should have the smallest possible estimate. Hence, from (1.3.8–1.3.10), the function $G(X)$ should be the solution of the extremum problems $\inf_G \int_{V_n} |G(X)| dV$, $\inf_G \int_{V_n} |G(X)|^2 dV$, or $\inf_G \|G(X)\|_C$.

The second principle for choosing $G(X)$ is that the integrand W in the DRE $\int_{\partial\Omega} W dS$ should possess few terms and not be very complicated. Finally, the DRE should have the highest possible degree of algebraic precision.

In general, the minimum estimate of ρ_m can be found only for some special regions (see next section for details). However, while we focus on the last two principles, we should still pay some attention to the first.

We now give a choice of $G(X)$ such that the corresponding m-term DRE possesses degree $2m-1$ of algebraic precision.

In formula (1.3.7), assume that Φ_1 and Φ_2 are smooth functions and V_{n-1} is a bounded and simply connected domain, then $G(X)$ can be chosen as

$$\begin{aligned} G(X) &= \frac{m!}{(2m)!} \left(\frac{\partial}{\partial x_n}\right)^m \{(x_n - \Phi_1(x_1, \dots, x_{n-1}))^m \\ &\quad \times (x_n - \Phi_2(x_1, \dots, x_{n-1}))^m\}. \end{aligned} \tag{1.3.11}$$

Expanding the above in terms of x_n, we can show that it is of the form given by (1.3.3).

Theorem 1.3.1 *Let $F(X) \in C^{2m}(V_n)$. DRE (1.3.7) with the $G(X)$ shown in (1.3.11) possesses an algebraic precision of degree $2m-1$.*

Proof. Take the remainder

$$\begin{aligned}\rho_m &= \frac{(-1)^m}{m!}\int_{V_n}\frac{\partial^m F}{\partial x_n^m}G(X)dV\\ &= \frac{(-1)^m}{(2m)!}\int_{V_n}\frac{\partial^m F}{\partial x_n^m}\frac{\partial^m}{\partial x_n^m}\{(x_n-\Phi_1(x_1,\dots,x_{n-1}))^m\\ &\quad\times(x_n-\Phi_2(x_1,\dots,x_{n-1}))^m\}dV,\end{aligned}$$

where $\partial^m F/\partial x_n^m$ and $\{(x_n-\Phi_1(x_1,\dots,x_{n-1}))^m(x_n-\Phi_2(x_1,\dots,x_{n-1}))^m\}$ are the F and G in formula (1.2.3), respectively. Applying the generalized integration by parts formula (1.2.3) with $L=\partial^m/\partial x_n^m$ to ρ, we obtain the corresponding dimensionality reducing expression terms

$$\int_{S_{n-1}} WdS = 0.$$

Therefore,

$$\begin{aligned}\rho_m &= \frac{1}{(2m)!}\int_{V_n}\frac{\partial^{2m} F}{\partial x_n^{2m}}\{(x_n-\Phi_1(x_1,\dots,x_{n-1}))^m\\ &\quad\times(x_n-\Phi_2(x_1,\dots,x_{n-1}))^m\}dV.\end{aligned}\tag{1.3.12}$$

It is obvious that $\rho_m=0$ for all polynomials F of degree less than or equal to $2m-1$. Hence, expression (1.3.7) is of the algebraic precision of degree $2m-1$ if $G(X)$ is defined as in (1.3.11). □

From inequalities (1.3.8)–(1.3.10), we immediately have the estimates

$$\begin{aligned}|\rho_m| &\le \frac{1}{(2m)!}\left\|\frac{\partial^{2m} F}{\partial x_n^{2m}}\right\|_{L_2}\|(x_n-\Phi_1(x_1,\dots,x_{n-1}))^m\\ &\quad\times(x_n-\Phi_2(x_1,\dots,x_{n-1}))^m\|_{L_2},\end{aligned}\tag{1.3.13}$$

$$\begin{aligned}|\rho_m| &\le \frac{1}{(2m)!}\left\|\frac{\partial^{2m} F}{\partial x_n^{2m}}\right\|_{C}\|(x_n-\Phi_1(x_1,\dots,x_{n-1}))^m\\ &\quad\times(x_n-\Phi_2(x_1,\dots,x_{n-1}))^m\|_{L_1},\end{aligned}\tag{1.3.14}$$

$$\begin{aligned}|\rho_m| &\le \frac{1}{(2m)!}\left\|\frac{\partial^{2m} F}{\partial x_n^{2m}}\right\|_{L_1}\|(x_n-\Phi_1(x_1,\dots,x_{n-1}))^m\\ &\quad\times(x_n-\Phi_2(x_1,\dots,x_{n-1}))^m\|_{C}.\end{aligned}\tag{1.3.15}$$

In addition, estimate (1.3.13) can be modified as follows.

Theorem 1.3.2 *Let $F(X) \in C^{2m}(V_n)$. DRE (1.3.7) with auxiliary function $G(X)$ defined as (1.3.11) has the following estimate for its remainder ρ_m.*

$$|\rho_m| \leq \left[\frac{1}{(4m)!}\right]^{\frac{1}{2}} \min_{i=1,2} \left\|(x_n - \Phi_i(x_1, \dots, x_{n-1}))^{2m}\right\|_{L_2} \left\|\frac{\partial^{2m} F}{\partial x_n^{2m}}\right\|_{L_2}. \quad (1.3.16)$$

Proof. Applying the generalized integration by parts formula (1.2.3) to the integral in the last factor on the right-hand of (1.3.13), we have

$$\begin{aligned}
&\int_{V_n} (x_n - \Phi_1(x_1, \dots, x_{n-1}))^{2m} (x_n - \Phi_2(x_1, \dots, x_{n-1}))^{2m}\, dV \\
= &\int_{V_n} (x_n - \Phi_1(x_1, \dots, x_{n-1}))^{2m} \\
&\times \frac{(2m)!}{(4m)!} \frac{\partial^{2m}}{\partial x_n^{2m}} (x_n - \Phi_1(x_1, \dots, x_{n-1}))^{4m}\, dV \\
= &\frac{(2m)!}{(4m)!}(2m)! \int_{V_n} (x_n - \Phi_1(x_1, \dots, x_{n-1}))^{4m}\, dV.
\end{aligned}$$

Hence, from (1.3.13), we obtain

$$\begin{aligned}
&|\rho_m| \\
\leq &\frac{1}{(2m)!} \left\|\frac{\partial^{2m} F}{\partial x_n^{2m}}\right\|_{L_2} \left(\int_{V_n} \frac{(2m)!}{(4m)!}(2m)!(x_n - \Phi_1(x_1, \dots, x_{n-1}))^{4m} dV\right)^{\frac{1}{2}} \\
= &\left[\frac{1}{(4m)!}\right]^{\frac{1}{2}} \|(x_n - \Phi_1(x_1, \dots, x_{n-1}))^{2m}\|_{L_2} \left\|\frac{\partial^{2m} F}{\partial x_n^{2m}}\right\|_{L_2}.
\end{aligned}$$

Similarly, we find

$$|\rho_m| \leq \left[\frac{1}{(4m)!}\right]^{\frac{1}{2}} \left\|(x_n - \Phi_2(x_1, \dots, x_{n-1}))^{2m}\right\|_{L_2} \left\|\frac{\partial^{2m} F}{\partial x_n^{2m}}\right\|_{L_2}.$$

The proof of Theorem 1.3.2 is thus complete. □

We will give more accurate estimates for the remainders of DREs over certain domains in the next section. In the remaining part of this section, we will discuss the construction of DREs with remainders that have an algebraic precision of degree $2m - 1$ over unit sphere, simplexes, cylindrical domains, etc.

Drawing inspiration from Gröbner's work in [23], we choose the following *Hermite–Didon polynomial* (1.3.17) and generalized *Appell polynomial* (1.3.18) as auxiliary polynomials $G(X)$ for estimating ρ_m of DREs of integrals over the unit sphere $B_n(x_1^2 + \cdots + x_n^2 \leq 1)$ and the simplex $T_n(x_i \geq 0, x_1 + \cdots + x_n \leq 1)$, respectively.

$$U_m(X) = \frac{m!}{(2m)!}\left(\frac{\partial}{\partial x_n}\right)^n (x_1^2 + \cdots + x_n^2 - 1)^m, \tag{1.3.17}$$

$$Q_m(X) = \frac{m!}{(2m)!}\left(\frac{\partial}{\partial x_n}\right)^n [x_n^m (x_1 + \cdots + x_n - 1)^m]. \tag{1.3.18}$$

Theorems 1.3.3 and 1.3.4 show that the corresponding remainder estimates are the smallest possible in a certain sense and the corresponding DREs possess m terms and degree $2m - 1$ of algebraic precision.

Theorem 1.3.3 *Suppose that $F(X)$ is a continuous function defined on the sphere $B_n(x_1^2+\cdots+x_n^2 \leq 1)$ that has $2m$-order continuous partial derivative with respect to x_n. Then there exists the following DRE that has m terms and possesses degree $2m - 1$ of algebraic precision.*

$$\int_{B_n} F(X)dV = \sum_{k=0}^{m-1} \frac{(-1)^k}{m!} \int_{S_{n-1}} L_k\left(F(X), U_m(X)\right) dS + \rho_m, \tag{1.3.19}$$

where $L_k(\cdot, \cdot)$ is defined by (1.3.4) and ρ_m has estimate

$$|\rho_m| \leq \frac{\pi^{\frac{n}{2}} \cdot m!}{\Gamma\left(m + \frac{n}{2} + 1\right)(2m)!} \left\| \frac{\partial^{2m} F}{\partial x_n^{2m}} \right\|_C \tag{1.3.20}$$

or

$$|\rho_m| \leq \left(\frac{\pi^{\frac{n}{2}} \cdot m!}{\Gamma\left(m + \frac{n}{2} + 1\right)(2m)!}\right)^{\frac{1}{2}} \left\| \frac{\partial^{2m} F}{\partial x_n^{2m}} \right\|_{L_2}. \tag{1.3.21}$$

Proof. From inequalities (1.3.12) and (1.3.13) and

$$\int_{B_n} \left(1 - \sum_{i=1}^n x_i^2\right)^m dV = \frac{\pi^{\frac{n}{2}} m!}{\Gamma\left(m + \frac{n}{2} + 1\right)}$$

we know that Theorem 1.3.3 is true. □

Theorem 1.3.4 *Let $F(X)$ be a continuous function defined on the simplex $T_n(x_i \geq 0, x_1 + \cdots + x_n \leq 1)$ that has $2m$-order continuous partial derivative with respect to x_n. Then there exists the following DRE that has m terms and possesses degree $2m - 1$ of algebraic precision.*

$$\int_{T_n} F(X)dV = \sum_{k=0}^{m-1} \frac{(-1)^k}{m!} \int_{S_{n-1}} L_k\left(F(X), Q_m(X)\right)dS + \rho_m, \tag{1.3.22}$$

where $L_k(\cdot,\cdot)$ is defined by (1.3.4) and ρ_m has estimate

$$|\rho_m| \le \frac{m!m!}{(2m+n)!(2m)!} \left\| \frac{\partial^{2m} F}{\partial x_n^{2m}} \right\|_C \tag{1.3.23}$$

or

$$|\rho_m| \le \left(\frac{1}{(4m+n)!} \right)^{\frac{1}{2}} \left\| \frac{\partial^{2m} F}{\partial x_n^{2m}} \right\|_{L_2}. \tag{1.3.24}$$

Proof. From inequalities (1.3.12) and (1.3.13) and

$$\int_{T_n} x_n^m \left(1 - \sum_{i=1}^n x_i\right)^m dV = \frac{m!m!}{\Gamma(2m+n+1)}$$

we know that Theorem 1.3.4 is true. □

In general, suppose that the boundary surface S_{n-1} of V_n is given by the equation

$$\Phi(x_1, \ldots, x_n) \equiv x_n^2 + gx_n + h = 0$$

and possibly some planes, $x_i = 0$, $i = 1, \ldots, n$. Here $g = g(x_1, \ldots, x_n)$ and $h = h(x_1, \ldots, x_n)$ are piecewise differentiable functions, and all points in V_n satisfy $\Phi \le 0$. Then, define auxiliary function $G(X)$ as

$$G(X) = \frac{m!}{(2m)!} \left(\frac{\partial}{\partial x_n} \right)^m (\Phi(x_1, \ldots, x_n))^m. \tag{1.3.25}$$

Similar to Theorems 1.3.3 and 1.3.4, we have the following result.

Theorem 1.3.5 *Let $G(X)$ be defined by (1.3.25); $\Phi = 0$ be the boundary surface equation of V_n. Assume that all points in V_n satisfy $\Phi \le 0$. Then DRE (1.3.4) possesses degree $2m-1$ of algebraic precision and has the remainder estimate*

$$|\rho_m| \le \frac{1}{(2m)!} \left\| \frac{\partial^{2m} F}{\partial x_n^{2m}} \right\|_{L_2} \left\| (\Phi(x_1, \ldots, x_n))^m \right\|_{L_2} \tag{1.3.26}$$

or

$$|\rho_m| \le \frac{1}{(2m)!} \left\| \frac{\partial^{2m} F}{\partial x_n^{2m}} \right\|_C \left\| (\Phi(x_1, \ldots, x_n))^m \right\|_{L_1}. \tag{1.3.27}$$

Example 1.3.1. As in the first example, we consider the half sphere $\Omega(0 \le z \le \sqrt{r^2 - x^2 - y^2})$ and choose the auxiliary function

$$G(X) = \frac{m!}{(2m)!}\left(\frac{\partial}{\partial z}\right)^m \left(z^m \cdot \left(z - \sqrt{r^2 - x^2 - y^2}\right)^m\right).$$

Consequently, we obtain the following m term DRE with an algebraic precision of degree $2m - 1$.

$$\int_\Omega F(x, y, z)dxdydz \approx \frac{1}{m!}\sum_{k=0}^{m-1}(-1)^k \int_{S_2} \frac{\partial^k F}{\partial z^k}\frac{\partial^{m-k-1} G}{\partial z^{m-k-1}} dxdy. \tag{1.3.28}$$

Its remainder has estimate

$$|\rho_m| \le \left(\frac{2\pi r^{4m+3}}{(4m+3)(4m+1)!}\right)^{\frac{1}{2}} \left\|\frac{\partial^{2m} F}{\partial z^{2m}}\right\|_{L_2}.$$

Example 1.3.2. For the cylindrical domain $\Omega(x^2 + z^2 \le r^2, a \le y \le b)$, we may choose the auxiliary function

$$G(X) = \frac{m!}{(2m)!}\left(\frac{\partial}{\partial z}\right)^m \left(x^2 + z^2 - r^2\right)^m$$

and obtain an m term DRE of algebraic precision degree $2m - 1$ over Ω, with the same form as formula (1.3.28) and the remainder estimate

$$|\rho_m| \le \frac{1}{(2m)!}\left(\frac{\pi(b-a)r^{4m+2}}{2m+1}\right)^{\frac{1}{2}} \left\|\frac{\partial^{2m} F}{\partial z^{2m}}\right\|_{L_2}.$$

1.4 Minimum estimation of remainders in DREs with algebraic precision

In Section 1.3, by a suitable choice of $G(x)$ we can make use of (1.2.3) to derive an expansion of the integral J, with an error term $\int_\Omega G \cdot L(F)\, dX$, in terms of some boundary surface integrals. This is a DRE when the error term is dropped. The second step of the DRE method is to minimize the error term by finding an appropriate G. This actually leads to a minimization problem of the functional $\int_\Omega G \cdot L(F)\, dX$ subject to certain required conditions for $G(X)$. We will discuss this minimization problem for some special integral region Ω. Without loss of generality, in the following we assume $L = \Lambda = D^{(m_1,\dots,m_n)}$, $m_1+\cdots+m_n = m$. Thus (1.2.3) can be rewritten as

$$\int_\Omega F\Lambda(G)\,dX = \int_{\partial\Omega} W\,dS + (-1)^m \int_\Omega G\Lambda(F)\,dX, \tag{1.4.1}$$

where W is defined by (1.2.6) and the region $\Omega \subset \mathbb{R}$ is bounded and closed with the piecewise smooth boundary $\partial\Omega$. In particular, for $n = 2$, $\partial\Omega$ is a simple closed curve.

Obviously, $\Lambda(G)$ can be constant for certain types of polynomials, and it follows that (1.4.1) is a DRE of the integral $\int_\Omega F\,dX$ with the error term $(-1)^m \int_\Omega G\Lambda(F)\,dX$. Hence, we will define G as the following three types of polynomials so that the error term will be minimized. Denote $i = i_1+\cdots+i_n$ and $m = m_1 + \cdots + m_n$, and let $\sum'_{i_1,\dots,i_n} a_{i_1\dots i_n} x_1^{i_1}\dots x_n^{i_n}$ be the linear combination of all monomials $x_1^{i_1}\dots x_n^{i_n}$, $0 \le i \le m$, except the term $x_1^{m_1}\dots x_n^{m_n}$. Using these notations, we define

$$\begin{aligned} & G_m^{(m_1,\dots,m_n)} \\ = & \{P(x) = x_1^{m_1}\dots x_n^{m_n} + \sum_{i_1,\dots,i_n}{}' a_{i_1\dots i_n} x_1^{i_1}\dots x_n^{i_n}, \\ & 0 \le i \le m\}, \\ & K_m^{(m_1,\dots,m_n)} \\ = & \{P(x) = x_1^{m_1}\dots x_n^{m_n} + \sum_{i_1,\dots,i_n} a_{i_1\dots i_n} x_1^{i_1}\dots x_n^{i_n}, \\ & 0 \le i_1,\dots,i_n \le m-1, 0 \le i \le m-1\}, \end{aligned}$$

and

$$\begin{aligned} & H_m^{(m_1,\dots,m_n)} \\ = & \{P(x) = x_1^{m_1}\dots x_n^{m_n} + \textstyle\sum_{i_1,\dots,i_n} a_{i_1\dots i_n} x_1^{i_1}\dots x_n^{i_n}, \\ & 0 \le i_k \le m_k; k = 1,\dots,n; 0 \le i \le m-1\}. \end{aligned}$$

In addition, we use π_m^n to denote the set of all polynomials defined in $\mathbb{R}^n$ with total degree not more than m and, in particular, denote $H^m \equiv H_1^m$. Obviously,

$$\begin{aligned} \pi_{m_1}(x_1)\pi_{m_2}(x_2) \quad \cdots \quad & \pi_{m_n}(x_n) \subset H_m^{(m_1,\dots,m_n)} \subset K_m^{(m_1,\dots,m_n)} \\ & \subset G_m^{(m_1,\dots,m_n)} \subset \pi_m^n \subset C^m. \end{aligned}$$

Here, the left-hand side of the above expression is the Cartesian product of $\pi_{m_i}(x_i)$, $i = 1, 2, \dots, n$, and the right-hand side, C^m, is the class of n variable functions with mth order continuous mixed partial derivatives.

If $G \in G_m^{(m_1,\dots,m_n)}$, we have the approximate DRE

$$\int_\Omega F(X)\,dX \approx \frac{1}{m_1!\dots m_n!}\int_{\partial\Omega} W(X)\,dS, \tag{1.4.2}$$

where

$$\begin{aligned} W(X) &= \sum_{i=1}^{n}\sum_{k=0}^{m_i-1}(-1)^{m_1+\cdots+m_{i-1}+k}D^{(m_1,\ldots,m_{i-1},k,0,\ldots,0)}F(X) \\ &\cdot \quad D^{(0,\ldots,0m_i-k-1,m_{i+1},\ldots,m_n)}G(X)\frac{\partial x_i}{\partial \nu} \end{aligned} \quad (1.4.3)$$

and $\partial x_i/\partial \nu$ is the outer normal derivative of x_i on the surface $\partial\Omega$. In expression (1.4.2), the omitted error term or remainder is

$$\rho_m = \frac{(-1)^m}{m_1!\ldots m_n!}\int_{\Omega} G(X)\Lambda(F)(X)\,dX. \quad (1.4.4)$$

Obviously, DRE (1.4.2) possesses algebraic precision degree $m-1$. In fact, it is exact for any polynomial $F(X) = \sum'_{i_1,\ldots,i_n} a_{i_1\ldots i_n}x_1^{i_1}\ldots x_n^{i_n}$, $i \leq m$. In particular, it is exact for all $F(X) \in \pi^n_{m-1}$, the set of all polynomials defined in $\mathbb{R}^n$ with total degree not more than $m-1$.

In the following theorems, we will give the optimal approximate DRE (1.4.2) in the sense that the remainder ρ_m possesses the minimum estimation. Similar to estimates (1.3.8)–(1.3.10), remainder (1.4.4) has the following three different estimates.

$$|\rho_m| \leq \frac{1}{m_1!\ldots m_n!}||\Lambda F||_{L_1}||G||_C, \quad (1.4.5)$$

$$|\rho_m| \leq \frac{1}{m_1!\ldots m_n!}||\Lambda F||_C||G||_{L_1}, \quad (1.4.6)$$

and

$$|\rho_m| \leq \frac{1}{m_1!\ldots m_n!}||\Lambda F||_{L_2}||G||_{L_2}. \quad (1.4.7)$$

Therefore, to construct the minimum estimate of ρ_m, we only need to find the extremal functions (polynomials), $G(X)$, in certain polynomial classes to obtain the minimum estimates of $||G||_C$, $||G||_{L_2}$, and $||G||_{L_1}$. The DREs associated with such $G(X)$'s are the optimal DREs, and they will be used to construct the optimal BTQFs with certain degrees of algebraic precision (see examples at the end of this section and in Chapter 3).

For a general higher dimensional region Ω, it is very difficult to find the corresponding *best Chebyshev approximation polynomial* or minimax approximation polynomial. However, there are several well-known minimax approximation polynomials on the n dimensional cube $V_n = \{(x_1,\ldots,x_n) : -1 \leq x_i \leq 1, i = 1,\ldots,n\}$, triangular region $\Delta(x \geq -1, y \geq -1, x+y \leq 0)$, and the unit disc R $(x^2+y^2 \leq 1)$. Hence, substituting the polynomials in (1.4.5), we immediately obtain the following results.

Theorem 1.4.1 *Suppose* $\Omega = V_n = \{(x_1, \ldots, x_n) : -1 \leq x_i \leq 1, i = 1, \ldots, n\}$ *is an n-dimensional cube. Let* $F(X) \in C^{(m_1,\ldots,m_n)}(\Omega)$*, the set of all functions defined on* Ω *that have continuous partial derivatives* $\partial^m F / \partial x_1^{m_1} \cdots \partial x_n^{m_n}$*, and let* $G(X) \in K_m^{(m_1,\ldots,m_n)}$*. Then the approximate DRE (1.4.2) is optimal if*

$$G(X) = G^* = \prod_{i=1}^{n} \prod_{k=1}^{m_i} (x_i + \cos \frac{2k-1}{2m_i} \pi) \tag{1.4.8}$$

and the minimum estimate for ρ_m *is*

$$|\rho_m| \leq \frac{1}{2^{m-n} m_1! \ldots m_n!} ||\Lambda F||_{L_1}. \tag{1.4.9}$$

Proof. It is sufficient to note that $G(X)$ in (1.4.8) is the best Chebyshev approximation polynomial or the minimax approximation polynomial; i.e., it satisfies $\|G^*\|_C = 2^{-m+n} = \min_{G \in K_m^{(m_1,\ldots,m_n)}} ||G(X)||_C$, $X \in V_s$. $\square$

Theorem 1.4.2 *Suppose* $G(x, y) \in k_2^{(m_1,m_2)}$*. Let* Ω_1 *and* Ω_2 *be the triangular region* $\Delta(x \geq -1, y \geq -1, x + y \leq 0)$ *and the unit disc* R $(x^2 + y^2 \leq 1)$*, respectively. Then the DREs over* Δ *and* R *are optimal if*

$$\begin{aligned} G_1(x, y) &= B_{m_1 m_2}(x, y) = \textstyle\prod_{i=1}^{m_1} (x + \cos \frac{2i-1}{2(m_1+m_2)} \pi) \\ &\quad \textstyle\prod_{j=1}^{m_2} (y + \cos \frac{2j-1}{2(m_1+m_2)} \pi) \end{aligned}$$

and

$$\begin{aligned} G_2(x, y) &= I_{m_1 m_2}(x, y) = \textstyle\prod_{i=1}^{m_1} (x + \sin \frac{2i-m_1-1}{2(m_1+m_2)} \pi) \\ &\quad \textstyle\prod_{j=1}^{m_2} (y + \sin \frac{2j-m_2-1}{2(m_1+m_2)} \pi), \end{aligned}$$

respectively. Here, the polynomials $B_{m_1 m_2}(x, y)$ *and* $I_{m_1,m_2}(x, y)$ *are the best Chebyshev approximation polynomials or the minimax approximation polynomials defined on* Δ *and* R*, respectively. The corresponding optimal DREs can be written as*

$$\begin{aligned} &\int_{\Omega_i} F(x, y) dx dy \\ \approx\ & \frac{1}{m_1! m_2!} \int_{\partial \Omega_i} \left\{ \sum_{k=0}^{m_1-1} (-1)^k D^{(k,0)} F(x, y) D^{(m_1-k-1,m_2)} G_i(x, y) \right\} dy \\ +\ & \frac{1}{m_1! m_2!} \int_{\partial \Omega_i} \left\{ \sum_{k=0}^{m_2-1} (-1)^{m_1+k+1} D^{(m_1,k)} F(x, y) \right. \\ & \left. D^{(0,m_2-k-1)} G_i(x, y) \right\} dx, \end{aligned} \tag{1.4.10}$$

where $i = 1, 2$. The optimal estimates for the above expansions are

$$|\rho_m| \le \left\| D^{(m_1,m_2)} F \right\|_{L_1(\Delta)} / [2^{m_1+m_2-1} m_1! m_2!]$$

and

$$|\rho_m| \le \left\| D^{(m_1,m_2)} F \right\|_{L_1(R)} / [2^{m_1+m_2-1} m_1! m_2!],$$

respectively.

From the existence theorem of orthogonal polynomials, we immediately have the following result.

Theorem 1.4.3 *There exists a unique $G^*(X) \in G_m^{(m_1,\dots,m_n)}$ with $\langle G^*, G^* \rangle = \min_{G \in G_m^{(m_1,\dots,m_n)}} \langle G, G \rangle$. Moreover, $G^*(X)$ is an orthogonal polynomial of degree m. The corresponding approximate DRE (1.4.2) is optimal for all $G \in G_m^{(m_1,\dots,m_n)}$ and the corresponding minimum estimate for the remainder is*

$$|\rho_m| \le \frac{1}{m_1! \dots m_n!} ||\Lambda F||_{L_2} \left\| G^* \right\|_{L_2}. \tag{1.4.11}$$

For simplification, we will show the existence of G^* for the case of $n = 2$, $m_1 = 2$, $m_2 = 1$. A general proof proceeds along similar lines.

Assume $G(x, y) \in P_2^{(2,1)}$; i.e.,

$$\begin{aligned} G(x, y) &= x^2 y + a_{00} + a_{10}x + a_{01}y + a_{20}x^2 + a_{11}xy + a_{02}y^2 \\ &\quad + a_{30}x^3 + a_{12}xy^2 + a_{03}y^3. \end{aligned}$$

Denote

$$\mu_{i,j} = \int_\Omega x^i y^j dx dy.$$

To prove the existence of G^*, we only need to prove that the matrix with the following column vector as its row entries is of full rank.

$$\begin{aligned} &(\mu_{i,j}, \mu_{i+1,j}, \mu_{i,j+1}, \mu_{i+2,j}, \mu_{i+1,j+1}, \mu_{i,j+2}, \\ &\mu_{i+3,j}, \mu_{i+2,j+1}, \mu_{i+1,j+2}, \mu_{i,j+3})^T, \end{aligned}$$

where $0 \le i, j \le 3$, $i + j \le 3$, $(i, j) \ne (2, 1)$. If the matrix does not have a full rank, then there exist a_{ij}, $0 \le i + j \le 3$, $(i, j) \ne (2, 1)$, that are not all zeroes, such that

$$\int_\Omega Q_{2,1}(x, y) x^i y^j dx dy = 0, \quad 0 \le i + j \le 3, (i, j) \ne (2, 1),$$

where

$$\begin{aligned}Q_{2,1}(x,y) &= \alpha_{00}+\alpha_{10}x+\alpha_{01}y+\alpha_{20}x^2+\alpha_{11}xy+\alpha_{02}y^2\\&\quad+\alpha_{30}x^3+\alpha_{12}xy^2+\alpha_{03}y^3.\end{aligned}$$

Hence,

$$\int_\Omega \left[Q_{2,1}(x,y)\right]^2 dxdy = \sum_{i,j}\alpha_{ij}\int_\Omega Q_{2,1}(x,y)x^iy^j dxdy = 0$$

which contradicts $\int_\Omega \left[Q_{2,1}(x,y)\right]^2 dxdy > 0$. Therefore, the theorem is proved for the case of $n=2$, $m_1=2$, $m_2=1$.

As for $G(X)\in K_m^{(m_1,\dots,m_n)}$, it can be written as

$$G(X) = x_1^{m_1}\dots x_n^{m_n} - p(X),$$

where $p(X)\in\pi_{m-1}^n$. Assume that an orthonormal basis of π_{m-1}^n is $\{p_i(X), i=1,\dots,n\}$, then we have $n\le\left(\begin{array}{c}m+n-2\\n-1\end{array}\right)$ from the Jackson theorem. Therefore, from the best approximation theorem for the inner product space, we have

$$\langle G^*,G^*\rangle = \min_{G\in K_m^{(m_1,\dots,m_n)}}\langle G,G\rangle$$

if and only if

$$G^*(X) = x_1^{m_1}\dots x_n^{m_n} - \sum_{i=1}^n\langle x_1^{m_1}\dots x_n^{m_n}, p_i\rangle p_i, \tag{1.4.12}$$

Thus, we have Theorem 1.4.4.

Theorem 1.4.4 *Suppose $F(X)\in C^{(m_1,\dots,m_n)}$ and $G(X)\in K_m^{(m_1,\dots,m_n)}$. Then the approximate DRE (1.4.2) is optimal if $G(X)=G^*(X)$, the function shown in (1.4.12). The corresponding remainder (1.4.4) has the minimum estimation shown in (1.4.11).*

For an axially symmetric region (i.e., a region that is symmetric about all axes), the corresponding orthogonal polynomial can be found easily. For example, let us consider a 2-dimensional axially symmetric domain and denote $\mu_{p,q}=\int_\Omega x^py^qdxdy$. If $m=6$, then there exists the following set of orthogonal polynomials in $\pi_2^{m-1}=\pi_2^5$.

$$\begin{aligned}G_{5,0} &= x^5+a_5x^3+b_5xy^2+c_5x,\\G_{4,1} &= x^4y+a_4x^2y+b_4y^3+c_4y,\\G_{3,2} &= x^3y^2+a_3x^3+b_3xy^2+c_3x,\\G_{2,3} &= x^2y^3+a_2x^2y+b_2y^3+c_2y,\\G_{1,4} &= xy^4+a_1x^3+b_1xy^2+c_1x,\\G_{0,5} &= y^5+a_0x^2y+b_0y^3+c_0y,\end{aligned}$$

where coefficients a_i, b_i, and c_i are determined by

$$\begin{pmatrix} \mu_{4,0} & \mu_{2,2} & \mu_{2,0} \\ \mu_{6,0} & \mu_{4,2} & \mu_{4,0} \\ \mu_{4,2} & \mu_{2,4} & \mu_{2,2} \end{pmatrix} \begin{pmatrix} a_i \\ b_i \\ c_i \end{pmatrix} = - \begin{pmatrix} \mu_{i+1,5-i} \\ \mu_{i+3,5-i} \\ \mu_{i+1,7-i} \end{pmatrix} \qquad i = 5, 3, 1,$$

$$\begin{pmatrix} \mu_{2,2} & \mu_{0,4} & \mu_{0,2} \\ \mu_{4,2} & \mu_{2,4} & \mu_{2,2} \\ \mu_{2,4} & \mu_{0,6} & \mu_{0,4} \end{pmatrix} \begin{pmatrix} a_i \\ b_i \\ c_i \end{pmatrix} = - \begin{pmatrix} \mu_{i,6-i} \\ \mu_{i+2,6-i} \\ \mu_{i,8-i} \end{pmatrix} \qquad i = 4, 2, 0.$$

Normalizing the obtained orthogonal polynomials produces the orthonormal polynomials $\{G_1, G_2, G_3, G_4, G_5, G_6\}$. Hence, for all $G \in K_2^{(m_1,m_2)}$, $m_1 + m_2 = m = 6$, the corresponding DRE remainder (1.4.4) possesses the minimum estimate

$$|\rho_6| \leq \frac{1}{m_1!m_2!} \left\| D^{(m_1,m_2)} F \right\|_{L_2} \left\| x^{m_1} y^{m_2} - \sum_{i=1}^{6} \langle x^{m_1} y^{m_2}, G_i \rangle G_i \right\|_{L_2} .$$

Obviously, the estimation of remainder (1.4.4) of DRE (1.4.2) also depends on the class of function F. For instance, because remainder (1.4.4) is zero for any $F \in \pi_n^{m-1}$, we immediately have

Proposition 1.4.5 *DRE (1.4.2) holds exactly for all $F \in \pi_n^{m-1}$.*

We now discuss the optimal estimation when $G \in H_m^{(m_1,\dots,m_n)}$ and $F \in C^{(m_1,\dots,m_2)}$.

Theorem 1.4.6 *Suppose $\Omega = V_s = [-1, 1]^s$, an s-dimensional cube; $F(X) \in C^{(m_1,\dots,m_n)}(\Omega)$; and $G(X) \in H_m^{(m_1,\dots,m_n)}$. Then the approximate DRE (1.4.2) is optimal if*

$$G(X) = \prod_{i=1}^{s} Q_{m_i}(x_i), \tag{1.4.13}$$

where

$$Q_m(x) = \frac{\sin((m+1)\cos^{-1} x)}{2^m \sqrt{1 - x^2}} \tag{1.4.14}$$

is the second kind of Chebyshev polynomial. *The corresponding minimum estimate of the remainder is*

$$|\rho_m| \leq \frac{1}{2^{m-n} m_1! \dots m_n!} ||\Lambda F||_C. \tag{1.4.15}$$

In addition, denoting $A = \{G(X) : G(x) = \prod_{i=1}^n g_i(x_i),\ g_i(x_i) \in \pi_{m_i}^1,\ i = 1, 2, \dots, n\}$, and $D_M = \{F(X) : F(X) \in C^{(m_1,\dots,m_n)}(V_n)$ and $|\Lambda F| \leq M$ for some $M\}$, we have

$$\inf_{G \in A} \sup_{F \in B} |\rho_m(F, G)| = \frac{M}{2^{m-n} m_1! \dots m_n!}. \tag{1.4.16}$$

Proof. Actually this is a consequence of the well-known Zolotarëff–Korkin theorem in a higher dimension (see [79]), which asserts that

$$\inf_{G\in A}\int_{V_n}|G(X)|\,dX=\inf_{G\in A}\prod_{i=1}^{n}\int_{-1}^{1}|Q_{m_i}(x_i)|\,dx_i=\prod_{i=1}^{n}\frac{1}{2^{m_i-1}}=\frac{1}{2^{m-n}}. \tag{1.4.17}$$

Moreover, it can be easily shown that

$$\sup_{F\in B}|\rho_m(F,G)|=\frac{M}{m_1!\dots m_n!}\int_{V_n}|G(X)|\,dX. \tag{1.4.18}$$

In fact, $|\rho_m(F,G)|$ is no more than the right-hand side of the above equation, and the optimal estimate can be achieved for $\Lambda^m F = M\cdot signG(X)$. Although this F is not in D_M, it can be approximated arbitrarily closely by a sequence in D_M. Therefore, equation (1.4.18) holds. Finally, (1.4.16) follows from (1.4.17) and (1.4.18). □

Using expansion (1.4.2) and the *Gaussian quadrature formula,* we can construct an optimal BTQF over a certain domain. Here an optimal BTQF is a quadrature formula with the highest possible degree of algebraic precision and the fewest possible nodes (evaluation points) on the boundary of the domain. To describe this concept clearly, we will show an example at the end of this section. A general theory and more examples will be given in the next chapter.

Let us consider the BTQF for the integral of $F:\mathbb{R}^n\to\mathbb{R}$, $F(X)=F(x_1,\dots,x_n)$, over $V_n=[-1,1]^n$. Assume $G(X)=\prod_{i=1}^n Q_{m_i}(x_i)$, where $Q_m(x)$ is the second kind of Chebyshev polynomial defined by (1.4.14). From (1.4.2) and (1.2.6) we have the expansion

$$\begin{aligned}\int_{V_n}F(X)\,dX&\approx\frac{1}{m_1!\dots m_n!}\sum_{i=1}^{n}\int_{V_{n-1}}[W(x_1,\dots,x_n)]_{x_i=-1}^{x_i=1}\,dS\\ &\approx\sum_{i=1}^{n}\int_{-1}^{1}\cdots\int_{-1}^{1}\frac{1}{m_1!\dots m_i!}\sum_{k=0}^{m_i-1}(-1)^{m_1+\dots+m_{i-1}+k}\\ &[D^{(m_1,\dots,m_{i-1},k,0,\dots,0)}FQ_{m_1}(x_1)\dots Q_{m_{i-1}}(x_{i-1})\\ &\frac{\partial^{m_i-k-1}}{\partial x_i^{m_i-k-1}}Q_{m_i}(x_i)]_{x_i=-1}^{x_i=1}\,dx_1\dots dx_{i-1}dx_{i+1}\dots dx_n.\end{aligned} \tag{1.4.19}$$

The remainder of the expansion, $\rho_m=\frac{(-1)^m}{m_1!\dots m_n!}\int_{V_n}G\Lambda(F)\,dX$, can be estimated by (1.4.15).

Clearly, expansion (1.4.19) holds exactly (i.e., $\rho_m=0$) for a polynomial of degree of at most $m-1$. Hence, we can only expect to construct BTQFs with algebraic precision degree of at most $m-1$.

On the other hand, if $F \in \pi^n_{m-1}$, the integrand of the integrals over V_{n-1} in the summation of equation (1.4.19) are polynomials of degree $m-1$ in $\mathbb{R}^{n-1}$. Therefore, each of these integrals can be expanded approximately to the summation of integrals over V_{n-2} by using (1.4.2), and these expansions are also exact for all $F \in \pi^n_{m-1}$.

By repeating this process of expansions, we can eventually and approximately expand the integral $\int_{V_n} F(X)\, dX$ as the summation of univariate integrals over $[-1, 1]$ (i.e., the integrals along $(x_1, \pm 1, \dots, \pm 1)$, $(\pm 1, x_2, \dots, \pm 1)$, $\dots$, $(\pm 1, \pm 1, \dots, x_n)$ with $-1 \le x_i \le 1$, $i = 1, 2, \dots, n$, respectively). And the expansion is exact if $F \in \pi^n_{m-1}$. For each univariate integral, let us choose a quadrature formula of algebraic precision degree $m-1$ with the least number of evaluation points. For instance, if $m = 2r$, we choose

$$\int_{-1}^{1} f(t)\, dt \approx \sum_{j=0}^{r-1} C_j f(t_j) \quad (-1 \le t_j \le 1),$$

where t_j, $j = 0, 1, \dots, r-1$, are the zeros of the nth degree *Legendre polynomial.* Hence, we obtain the following result.

Theorem 1.4.7 *There exists a kind of optimal BTQF for the integral $\int_{V_s} F(X)\, dX$ that possesses the highest possible algebraic precision degree, $2r-1$, and uses $2^{n-1}nr$ evaluation points on the "edges" of the cubical domain $[-1, 1]^n$ as follows: $(t_j, \pm 1, \dots, \pm 1)$, $(\pm 1, t_j, \dots, \pm 1)$, $\dots$, $(\pm 1, \dots, \pm 1, t_j)$, $j = 0, 1, \dots, r-1$, where t_j are the zeros of the nth degree Legendre polynomials.*

Chapter 2

Boundary Type Quadrature Formulas with Algebraic Precision

First, in Section 1, we will show that a DRE can be made an effective tool for constructing BTQFs with preassigned algebraic precision. Section 2 will discuss the construction of BTQFs with homogeneous precision. *General Hermite formulas* and a class of BTQFs that has the highest possible degrees of algebraic precision will also be given. In Section 3, we will show a general process to construct quadratures for boundary integrals by using periodic wavelets and the sampling theorem. Finally, several applications of DREs and BTQFs will be given in Section 5.

2.1 Construction of BTQFs using DREs

In this section, we will apply the DREs with algebraic precision to give a method for constructing BTQFs with algebraic precision. In addition, for some special domains, we will construct the optimal quadrature formulas; i.e., formulas that possess the required degrees of algebraic precision and the fewest possible nodes (evaluation points) on the boundary of the domain.

The key for constructing a BTQF from a DRE is to choose a suitable numerical quadrature formula with the same degree of algebraic precision as that of the DRE used. Hence, the obtained BTQF will have the same algebraic precision as that of the DRE.

In this section, we shall examine the 3-dimensional case, but all results can be extended to the n-dimensional case without much difficulty.

Define $\Omega \subset \mathbb{R}^3$ as

$$\Omega := \{(x, y, z) : \phi_1(x, y) \le z \le \phi_2(x, y), (x, y) \in E \subset \mathbb{R}^2\}, \tag{2.1.1}$$

where ϕ_1 and ϕ_2 are bivariate smooth functions and E a convex region with a piecewise smooth closed curve, ∂E, as boundary. Denote $X \equiv (x, y, z)$ and $dX \equiv dxdydz$, and assume that $F(X), G(X) \in C^m(\Omega)$, where m is a positive integer.

To construct a BTQF with a simpler structure, we may choose $\Lambda = \frac{1}{m!}\left(\frac{\partial}{\partial z}\right)^m$ and assume that $G(X)$ takes the form

$$G(X) = z^m + h(x, y)z^{m-1} + \text{lower degree terms in z}, \tag{2.1.2}$$

where $h(x, y)$ and the coefficients of all the lower degree terms in z may be any smooth function of (x, y). Then by using expression (1.2.3), we obtain

$$J \equiv \int_\Omega F(X)dX \approx \frac{1}{m!}\sum_{k=0}^{m-1}(-1)^k \int_{\partial\Omega}\left(\frac{\partial}{\partial z}\right)^k F \left(\frac{\partial}{\partial z}\right)^{m-k-1} G dxdy, \tag{2.1.3}$$

where the right-hand side is an expansion with the following remainder term dropped.

$$\rho_m = \frac{(-1)^m}{m!}\int_\Omega G(X)\left(\frac{\partial}{\partial z}\right)^m F dX. \tag{2.1.4}$$

A direct estimate of $|\rho_m|$ was obtained in (1.4.11), namely

$$|\rho_m| \le \frac{1}{m!}\left\|\frac{\partial^m F}{\partial z^m}\right\|_{L_2} ||G||_{L_2}.$$

Theorem 1.3.5 minimized the L_2-norm of G, $||G||_{L_2}$, by using an orthogonal basis of π^2_{m-1} over Ω. However, for general Ω, we can only approximately minimize $||G||_{L_2}$ under appropriate conditions. Gröbner [23] has shown that Appell polynomials and Hermite–Didon polynomials are just the solutions for minimizing $||G||_{L_2}$ over simplexes and spheres, respectively (see Section 1.3 and examples at the end of this section). Thus, inspired by Gröbner's result concerning the above-mentioned polynomials, one may make an appropriate choice for $G(X)$ in the following form (see also equation (1.3.11))

$$G(X) = \frac{m!}{(2m)!}\left(\frac{\partial}{\partial z}\right)^m [(z - \phi_1(x, y))^m (z - \phi_2(x, y))^m]. \tag{2.1.5}$$

Obviously, $G(X)$ is of form (2.1.2) and implies the special polynomials mentioned previously as particular cases. Moreover, it can be shown that such a choice of $G(X)$ can even yield an extension of the *Hermite formula.* We rewrite Theorems 1.3.1 and 1.3.2 in the following 3-dimensional setting.

Theorem 2.1.1 *Let $F(X) \in C^{2m}(\Omega)$ and denote $\phi_i \equiv \phi_i(x, y)$, $i = 1, 2$. Then we have the generalized Hermite formula*

$$\int_\Omega F(X)dX = \sum_{v=1}^{m} \frac{1}{v!} \frac{\binom{m}{v}}{\binom{2m}{v}} \int_E H_v(x, y)dxdy + \rho_m, \tag{2.1.6}$$

where $H_v(x, y)$ is defined by

$$H_v(x, y) = (\phi_2 - \phi_1)^v [F_z^{(v-1)}(x, y, \phi_1) + (-1)^{v-1} F_z^{(v-1)}(x, y, \phi_2)], \tag{2.1.7}$$

$F_z^{(\alpha)} = (\partial/\partial z)^\alpha F$*, and the remainder term is given by*

$$\rho_m = \frac{1}{(2m)!} \int_\Omega (z - \phi_1)^m (z - \phi_2)^m F_z^{(2m)} dX. \tag{2.1.8}$$

In addition, formula (2.1.6) is of the highest possible algebraic precision, $2m-1$, and ρ_m has the following estimate.

$$|\rho_m| \le \left(\frac{1}{(4m)!}\right)^{1/2} \mu \left\| F_z^{(2m)} \right\|_{L_2}, \tag{2.1.9}$$

where $\mu = \min\{\|(z - \phi_1)^{2m}\|_{L_2}, \|(z - \phi_2)^{2m}\|_{L_2}\}$.

Let us now choose a numerical integration formula of the form

$$\int_E f(x, y)dxdy \approx \sum_{i \in I} a_i f(x_i, y_i), \tag{2.1.10}$$

where a_i are the corresponding quadrature coefficients. Of course, the evaluation points to be used are assumed to be inside E, the region of integration. Thus, applying (2.1.10) to each of the integrals contained in the summation on the right-hand side of equation (2.1.6) while omitting ρ_m, we obtain

$$J \equiv \int_\Omega F(X)dX \approx \sum_{v=1}^{m} \frac{1}{v!} \frac{\binom{m}{v}}{\binom{2m}{v}} \sum_{i \in I} a_i H_v(x_i, y_i), \tag{2.1.11}$$

where the values $H_v(x_i, y_i)$ are computed by (2.1.7) with $(x, y) = (x_i, y_i)$.

Equation (2.1.11) is a 3-dimensional BTQF for computing J because all nodes $(x_i, y_i, \phi_1(x_i, y_i))$ and $(x_i, y_i, \phi_2(x_i, y_i))$ $(i \in I)$ are on the boundary of Ω. From Theorem 2.1.1, DRE (2.1.6) is of algebraic precision degree $2m - 1$. And numerical quadrature formula (2.1.10) is also of a certain degree of algebraic precision. It follows that the BTQF (2.1.11) possesses some degree of algebraic precision. However, in order to make formula (2.1.11) to be of the highest possible degree of algebraic precision, we need to determine the smallest possible degree of algebraic precision for the quadrature formula (2.1.10). We give the following result.

Theorem 2.1.2 *Suppose that Ω is a bounded domain defined by (2.1.1) in which $\phi_1(x, y)$ and $\phi_2(x, y)$ are bivariate polynomials with degrees $deg\, \phi_1 = r$ and $deg\, \phi_2 = s$. Then in order to have formula (2.1.11) possess the highest possible degree of algebraic precision, $2m - 1$, the formula (2.1.10) to be employed should be of algebraic precision degree of no less than t, which is given by*

$$t = \begin{cases} 2m - 1, & where\ r = s = 0; \\ 2m \cdot \max(r, s), & otherwise. \end{cases} \tag{2.1.12}$$

Proof. Since the expansion given by (2.1.6) without ρ_m is of the highest possible algebraic precision degree, $2m-1$, it suffices to consider the case where F is taken to be a polynomial in x, y, and z, with degree $deg\, F(x, y, z) = 2m - 1$. Also, we shall use the notation $(x)_+ = \max\{0, x\}$. Ignoring the numerical coefficient, a typical term of $F(x, y, z)$ may be written as $x^\alpha y^\beta z^\gamma$, where $\alpha + \beta + \gamma \le 2m - 1$. Now it is easy to compute the maximum of $deg\, H_v(x, y)$, $v = 1, \ldots, m$, as follows.

$$\max_{1\le\gamma\le m} \{deg\, H_v(x, y)\} = \max_{1\le v\le m} \{v \cdot \max(r, s) + \max_{\alpha+\beta+\gamma\le 2m-1} (\alpha + \beta + (\gamma - v + 1)_+ \cdot \max(r, s))\}.$$

Thus, when $r = s = 0$, $\max_{1\le\gamma\le m}\{deg\, H_v(x, y)\} = 2m - 1$; when $r + s \ge 1$, $\max_{1\le\gamma\le m}\{deg\, H_v(x, y)\} = 2m \cdot \max\{r, s\}$. Therefore, if the degree of algebraic precision of formula (2.1.10) is no less than t, then the formula holds exactly for $f(x, y) = H_v(x, y)$, $v = 1, \ldots, m$. Consequently (2.1.11) is a formula of algebraic precision $2m - 1$. Moreover, the degree of algebraic precision of (2.1.10) should not be less than t, because the value $t := \max\{deg\, H_v(x, y)\}$ can be precisely reached. Hence, the theorem is true. □

In the following, we will give examples of boundary quadrature formulas over a simplex, a cubical domain, a unit sphere, etc.

Example 2.1.1. For the simplex $\Omega = \{(x, y, z) : x + y + z \le a, x, y, z \ge 0\}$, we may express it in form (2.1.1) with $\phi_1 \equiv 0$, $\phi_2 \equiv a - x - y$, and $E \equiv \Delta :=$

$\{(x, y) : x + y \leq a, x, y \geq 0\}$. Accordingly, the required auxiliary function G takes the form

$$G(X) = \frac{m!}{(2m)!} \left(\frac{\partial}{\partial z}\right)^m [z^m (x + y + z - a)^m].$$

This is the well-known Appell polynomial in three variables. Notice that in the present case $r = deg\,\phi_1 = 0$ and $s = deg\,\phi_2 = 1$, so we should take $t = 2m$ from Theorem 2.1.2. Thus, if we choose any numerical integration formula (2.1.10) for the Δ-region with a degree of algebraic precision no less than $2m$, we can obtain a BTQF of form (2.1.11) with

$$\begin{aligned} H_v(x_i, y_i) &= (a - x_i - y_i)^v (F_z^{(v-1)}(x_i, y_i, 0) \\ &+ (-1)^{v-1} F_z^{(v-1)}(x_i, y_i, a - x_i - y_i)) \end{aligned}$$

and with the error term $|\rho_m| \leq \left(\frac{a^{4m+3}}{(4m+3)!}\right)^{1/2} \left\| F_z^{(2m)} \right\|_{L_2}$.

Example 2.1.2. For the cubical domain $\Omega = \{(x, y, z) : |x| \leq 1, |y| \leq 1, |z| \leq 1\}$, we have $\phi_1(x, y) \equiv -1$, $\phi_2(x, y) \equiv 1$, and $E \equiv \{(x, y) : |x| \leq 1, |y| \leq 1\}$. Accordingly we take

$$G(X) = \frac{m!}{(2m)!} \left(\frac{\partial}{\partial z}\right)^m (z^2 - 1)^m.$$

This is just the m-th degree Legendre polynomial with the leading term z^m. Since $r = s = 0$, we require $t = 2m - 1$. Similar to the example shown at the end of Section 1.4, we apply the *Gauss–Legendre m-point quadrature formula* to the boundary integral and obtain a boundary quadrature formula for $\int_\Omega F(X)dX$ of precision $2m - 1$:

$$\int_\Omega F(X)dX \approx \sum_{v=1}^{m} \frac{2^v}{v!} \frac{\binom{m}{v}}{\binom{2m}{v}} \sum_{i=0}^{m-1} \sum_{j=0}^{m-1} c_i c_j H_v(x_i, y_j),$$

where $H_v(x_i, y_j) = F_z^{(v-1)}(x_i, y_j, -1) + (-1)^{v-1} F_z^{(v-1)}(x_i, y_j, 1)$; c_i are the Gaussian quadrature coefficients; and x_i, y_j are the zeros of the m-th degree Legendre polynomial. Moreover, the corresponding error term to be dropped has the estimate

$$|\rho_m| \leq \left(\frac{2^{4m+3}}{(4m+1)!}\right)^{1/2} \left\| F_z^{(2m)} \right\|_{L_2}.$$

Here, none of the errors caused by using the Gaussian quadrature formula are included.

Example 2.1.3. Consider the domain $\Omega = \{(x, y, z) : 0 \le z \le r^2 - x^2 - y^2, x^2 + y^2 \le r^2\}$ bounded by the quadratic surface $z = r^2 - x^2 - y^2$ and the circular disc $E \equiv \{(x, y) : x^2 + y^2 \le r^2\}$. We have $\phi_1(x, y) \equiv 0$ and $\phi_2(x, y) \equiv r^2 - x^2 - y^2$. Accordingly $G(X)$ should take the form

$$G(X) = \frac{m!}{(2m)!} \left(\frac{\partial}{\partial z}\right)^m [z^m (z + x^2 + y^2 - r^2)^m].$$

Since $deg\, \phi_1 = 0$ and $deg\, \phi_2 = 2$, $t = 4m$. This means that we have to employ a numerical integration formula (2.1.10) of algebraic precision of at least $4m$ for integrals over the circular region E. And consequently we can get a BTQF.

$$\int_\Omega F(X)dX$$
$$\approx \sum_{v=1}^{m} \frac{1}{v!} \frac{\binom{m}{v}}{\binom{2m}{v}} \sum_{i \in I} a_i (r^2 - x_i^2 - y_i^2)^v \times$$
$$\left(F_z^{(v-1)}(x_i, y_i, 0) + (-1)^{v-1} F_z^{(v-1)}(x_i, y_i, r^2 - x_i^2 - y_i^2)\right),$$

which possesses algebraic precision degree $2m - 1$.

Example 2.1.4. For the unit sphere $B_3 = \{(x, y, z) : x^2 + y^2 + z^2 \le 1\}$, we have $\phi_1(x, y) \equiv -\sqrt{1 - x^2 - y^2}$, $\phi_2(x, y) \equiv \sqrt{1 - x^2 - y^2}$, and $E \equiv \{(x, y) : x^2 + y^2 \le 1\}$. $G(X)$ takes the form

$$G(X) = \frac{m!}{(2m)!} \left(\frac{\partial}{\partial z}\right)^m (x^2 + y^2 + z^2 - 1)^m.$$

This is just the Hermite–Didon polynomial in 3-dimension. But now ϕ_1 and ϕ_2 are not polynomial functions. Certainly one cannot normally expect to use the generalized Hermite formula to obtain BTQFs of form (2.1.11) with algebraic precision. However, in this case one can directly make use of (2.1.3), thus obtaining

$$\int_{B_3} F(X)dX \approx$$
$$\sum_{k=0}^{m-1} \frac{(-1)^k}{(2m)!} \int_{\bar{S}_3} F_z^{(k)} [(x^2 + y^2 + z^2 - 1)^m]_z^{(2m-k-1)} z dS, \quad (2.1.13)$$

where $\bar{S}_3$ is the surface of the unit sphere B_3.

To see what precision can be attained, we may take F to be a polynomial in a single variable z with $deg\, F \ge m - 1$. Clearly, the degree of the integrand of each integral in the sum of (2.1.13) is $(deg\, F - k) + k + 2 = deg\, F + 2$. Because of (2.1.4) and (2.1.8) we may assume that $deg\, F = m - 1$ or $deg\, F = 2m - 1$, so that the integrand in each boundary integral will have degree $m+1$ or $2m+1$. To integral J in (2.1.13), we apply fully symmetric integration formulas for spherical surface integrals (see Keast and Diaz [49] or Section 2.5) with algebraic precision degree $m+1$ or $2m+1$. We thus obtain a BTQF with algebraic precision $m-1$ or $2m-1$, respectively.

2.2 BTQFs with homogeneous precision

In this section we shall develop a type of *BTQF with homogeneous precision* for integrals of the form

$$\int_\Omega F(x)dX = \int_a^b dx \int_{\psi_1(x)}^{\psi_2(x)} dy \int_{\phi_1(x,y)}^{\phi_2(x,y)} f(X)dz, \tag{2.2.1}$$

where Ω is a bounded closed domain and $X = (x, y, z)$. In addition, ψ_i and ϕ_i, $i = 1, 2$, are smooth functions such that $\phi_1(x, y) < \phi_2(x, y)$ for $a \le b$ and $\psi_1(x) \le y \le \psi_2(x)$. Evidently, (2.2.1) can be transformed into a triple integral taken over the cube $[-1, 1]^3$ by using simply the successive substitutions:

$$\begin{aligned} z &= \phi_1(x,y) + \tfrac{1}{2}(\phi_2(x,y) - \phi_1(x,y))(1+z), \\ y &= \psi_1(x) + \tfrac{1}{2}(\psi_2(x) - \psi_1(x))(1+y), \\ x &= a + \tfrac{1}{2}(b-a)(1+x). \end{aligned}$$

Here the induced mapping may be written as $T : X \to X = (x, y, z)$, though which $\Omega \to [-1, 1]^3$ and $f(X) \to T(f)(X)$, and consequently we may write

$$\int_\Omega f(x)dX = \int_{[-1,1]^3} T(f)(X)J(x,y)dxdydz, \tag{2.2.2}$$

where $J(x, y) = \partial(x, y, z)/\partial(x, y, z)$ is the Jacobian of T, given by

$$\frac{1}{8}(b-a)(\psi_2(x) - \psi_1(x))(\phi_2(x,y) - \phi_1(x,y)) \to J(x,y),$$

so that $J(x, y) > 0$ everywhere.

Thus, in the first place we have to confine ourselves to finding *symmetric BTQFs* for the simpler integrals

$$I(f) = \int_Q f(X^2)dX^2, \qquad Q \equiv [-1,1]^2, \tag{2.2.3}$$

and

$$J(F) = \int_K F(X^3) dX^3, \qquad K \equiv [-1, 1]^3, \tag{2.2.4}$$

where $X^2 \equiv (x, y)$, $dX^2 \equiv dxdy$, $X^3 \equiv (x, y, z)$, and $dX^3 \equiv dxdydz$. What we need to use are the symmetric differential operators

$$\begin{aligned} \Delta_2 &\equiv \partial^2/\partial x \partial y, \qquad \Delta_3 \equiv \partial^3/\partial x \partial y \partial z, \\ \Lambda_2 &\equiv \left(\tfrac{1}{m!}\right)^2 \Delta_2^m, \quad \Lambda_3 \equiv \left(\tfrac{1}{m!}\right)^3 \Delta_3^m. \end{aligned}$$

Because of the above differential operators, we choose auxiliary functions (polynomials) of the forms

$$g(X^2) = x^m y^m + R(x, y) \tag{2.2.5}$$

and

$$G(X^3) = x^m y^m z^m + S(x, y, z), \tag{2.2.6}$$

where the degrees of R and S are less than those of $x^m y^m$ and $x^m y^m z^m$ respectively; in other words, $deg_x R \leq m$, $deg_y R \leq m$ with $deg\ R \leq 2m - 1$, etc.

Observe that $f(X^2) = f(X^2)\Lambda_2 g(X^2)$ and $F(X^3) = F(X^3)\Lambda_3 G(X^3)$. Thus, assuming $\Delta_2^m f \in C(Q)$ and $\Delta_3^m F \in C(K)$, and applying the symmetric rule (1.2.3) to integrals (2.2.3) and (2.2.4), we obtain a pair of DREs as follows:

$$\begin{aligned} I(f) &\approx \frac{1}{(m!)^2} \int_{-1}^{1} \left[\sum_{k=0}^{m-1} \Delta_2^k g'_y \Delta_2^{m-k-1} f \right]_{x=-1}^{x=1} dy \\ &- \frac{1}{(m!)^2} \int_{-1}^{1} \left[\sum_{k=0}^{m-1} \Delta_2^k g \Delta_2^{m-k-1} f'_x \right]_{y=-1}^{y=1} dx, \end{aligned} \tag{2.2.7}$$

and

$$\begin{aligned} J(F) &\approx \frac{1}{(m!)^3} \sum_{k=0}^{m-1} (-1)^{3k} \int_{\partial K} \Big(\Delta_3^k F \Delta_3^{m-k-1} G''_{yz} dydz \\ &- \Delta_3^k F'_x \Delta_3^{m-k-1} G'_z dzdx + \Delta_3^k F''_{xy} \Delta_3^{m-k-1} G dxdy \Big), \end{aligned} \tag{2.2.8}$$

where the remainder terms that have been dropped are

$$\rho_m^1 \equiv \rho_m^1(f, g) = \frac{1}{(m!)^2} \int_Q g(X^2) \Delta_2^m f(X^2) dX^2, \tag{2.2.9}$$

and

$$\rho_m^2 \equiv \rho_m^1(F, G) = \frac{(-1)^{3m}}{(m!)^2} \int_K G(X^3)\Delta_3^m F(X^3) dX^3, \tag{2.2.10}$$

respectively.

What we are going to do is find appropriate g and G such that $|\rho_m^1|$ and $|\rho_m^2|$ have the smallest possible estimates (see Theorem 1.4.6). First define the following classes of functions with respect to a given constant $M > 0$:

$$\begin{aligned} E_2 &\equiv E_2(M) := \{f(X^2) : \Delta_2^m f \in C(Q), |\Delta_2^m f| \le M\}, \\ E_3 &\equiv E_3(M) := \{F(X^3) : \Delta_3^m F \in C(K), |\Delta_3^m F| \le M\}. \end{aligned}$$

Theorem 1.4.6 suggests that one may choose $g(X^2) = Q_m(x)Q_m(y)$ and $G(X^3) = Q_m(x)Q_m(y)Q_m(z)$ so that both DREs (2.2.7) and (2.2.8) become optimal DREs with respect to E_2 and E_3 respectively. Hence, we have the following result.

Theorem 2.2.1 *For any given $M > 0$ we have*

$$\inf_g \sup_{f \in E_2} |\rho_m^1(f, g)| = \frac{M}{4^{m-1}(m!)^2} \tag{2.2.11}$$

and

$$\inf_G \sup_{F \in E_3} |\rho_m^2(F, G)| = \frac{M}{8^{m-1}(m!)^3}, \tag{2.2.12}$$

where g and G range over all the polynomials of the form (2.2.5) and (2.2.6) respectively. Moreover, the infimums can be precisely attained with $g(X^2) = Q_m(x)Q_m(y)$ and $G(X^3) = Q_m(x)Q_m(y)Q_m(z)$. Here $Q_m(x)$ is the Chebyshev polynomial of the second kind, as shown in equation (1.4.14), with the leading term x^m,

$$Q_m(x) = \frac{\sin((m+1)\ \cos^{-1} x)}{2^m\sqrt{1-x^2}}.$$

In view of (2.2.9) it is clear that $\rho_m^1(f, g)$ vanishes for all bivariate polynomials of the form

$$f := f(x, y) = ax^m y^{m-1} + bx^{m-1}y^m + \text{lower degree terms}. \tag{2.2.13}$$

For brevity we may denote by $2 \times \langle m-1 \rangle + 1$ the homogeneous degree of (x, y). Similarly, the homogeneous degree of the polynomial

$$\begin{aligned} F := F(x, y, z) &= ax^m(yz)^{m-1} + by^m(xz)^{m-1} + cz^m(xy)^{m-1} \\ &\quad + \text{lower degree terms} \end{aligned} \tag{2.2.14}$$

may be denoted by $3 \times \langle m-1 \rangle + 1$. Clearly, F makes $\rho_m^2(f, g)$ vanish.

We may say that a BTQF is optimal if with respect to a given precision, the number of evaluation points (boundary points) to be used is the least possible, and moreover, the remainder term to be dropped from the DRE has some smallest possible estimate. What we now want to seek are formulas for $J(F)$ with the *homogeneous precision* degree $3 \times \langle m-1 \rangle + 1$ (i.e., the formulas are exact for all polynomials with form (2.2.14). We still need a lemma concerning $I(F)$ as a bridge leading to $J(F)$.

Lemma 2.2.2 *Suppose that*

$$\int_{-1}^{1} u(t)dt \approx \sum_{j=0}^{n-1} a_j u(t_j) \tag{2.2.15}$$

is a quadrature formula of given algebraic precision degree m with the least number of terms. Then we have an optimal BTQF for $I(f)$, with precision $2 \times \langle m-1 \rangle + 1$, of the form

$$I(f) \approx \frac{1}{(m!)^2} \sum_{k=0}^{m-1} \sum_{j=0}^{n-1} a_j \Phi(t_j), \tag{2.2.16}$$

where $\Phi(t)$ is given by

$$\begin{aligned} \Phi(t) \equiv\ & Q_m^{(k)}(1) \left\{ Q_m^{(k+1)}(t) \Delta_2^{m-k-1} f(1,t) - Q_m^{(k)}(t) \Delta_2^{m-k-1} f'_x(t,1) \right\} \\ & - Q_m^{(k)}(-1) \left\{ Q_m^{(k+1)}(t) \Delta_2^{m-k-1} f(-1,t) \right. \\ & \left. - Q_m^{(k)}(t) \Delta_2^{m-k-1} f'_x(t,-1) \right\}. \end{aligned} \tag{2.2.17}$$

Proof. Formula (2.2.16) with (2.2.17) can be readily obtained by applying (2.2.15) to each of the integrals on the right-hand side of (2.2.7) after replacing $g(X^2)$ by $Q_m(x)Q_m(y)$. To justify the precision $2 \times \langle m-1 \rangle + 1$ assume that $f(X^3)$ takes the form (2.2.13). Then (2.2.7) becomes exact since $\rho_m^1(f, g) = 0$ in this case. Moreover, we have

$$deg\left(\Delta_2^k g(x, \pm 1) \cdot \Delta_2^{m-k-1} f'_x(x, \pm 1)\right) = m$$

and

$$deg\left(\Delta_2^k g'_y(\pm 1, y) \cdot \Delta_2^{m-k-1} f(\pm 1, y)\right) = m$$

so that no error will be introduced on applying (2.2.15) to the right-hand side of (2.2.7). Hence (2.2.16) is exact for such $f(X^2)$, and it has the precision claimed. □

Now making use of the optimal expansion (2.2.8) with $G(X^3) = Q_m(x) Q_m(y)Q_m(z)$ we get

$$
\begin{aligned}
J(F) \approx\ & \frac{1}{(m!)^3}\sum_{k=0}^{m-1}(-1)^{3k}\left\{\int_{\partial K}[X_k(x,y,z)]_{x=-1}^{x=1}\,dydz\right. \\
& -\int_{\partial K}[Y_k(x,y,z)]_{y=-1}^{y=1}\,dzdx \\
& \left.+\int_{\partial K}[Z_k(x,y,z)]_{z=-1}^{z=1}\,dxdy\right\}, \qquad (2.2.18)
\end{aligned}
$$

where X_k, Y_k, and Z_k are given by

$$
\begin{aligned}
X_k(x,y,z) &= \Delta_3^k F(x,y,z)Q_m^{(m-k-1)}(x)Q_m^{(m-k)}(y)Q_m^{(m-k)}(z), \\
Y_k(x,y,z) &= \Delta_3^k F_x'(x,y,z)Q_m^{(m-k-1)}(x)Q_m^{(m-k-1)}(y)Q_m^{(m-k)}(z), \\
Z_k(x,y,z) &= \Delta_3^k F_{xy}''(x,y,z)Q_m^{(m-k-1)}(x)Q_m^{(m-k-1)}(y)Q_m^{(m-k-1)}(z).
\end{aligned}
$$

Clearly, formula (2.2.16) can be used to evaluate the integrals on the right-hand side of (2.2.18). Doing so, we obtain

$$
J(F) \approx \frac{1}{(m!)^5}\sum_{k=0}^{m-1}\sum_{i=0}^{m-1}\sum_{j=0}^{n-1}\Psi_{kij}(t_j), \qquad (2.2.19)
$$

where $\Psi_{kij}(t_j)$ is defined by

$$
\Psi_{kij}(t_j) = (-1)^{3k}a_j\left(\hat{X}_k^i(t_j) - \hat{Y}_k^i(t_j) + \hat{Z}_k^i(t_j)\right) \qquad (2.2.20)
$$

in which $\hat{X}_k^i(t_j)$ takes the form

$$
\begin{aligned}
\hat{X}_k^i(t_j) = &\left[\left[Q_m^{(i)}(y)Q_m^{(i+1)}(t_j)\left(\frac{\partial^2}{\partial y\partial z}\right)^{m-i-1}X_k(x,y,t_j)\right]_{y=-1}^{y=1}\right]_{x=-1}^{x=1} \\
-&\left[\left[Q_m^{(i)}(t_j)Q_m^{(i)}(z)\left(\frac{\partial^2}{\partial y\partial z}\right)^{m-i-1}\left(\frac{\partial}{\partial y}\right)X_k(x,t_j,z)\right]_{z=-1}^{z=1}\right]_{x=-1}^{x=1}.
\end{aligned}
$$

$\hat{Y}_k^i(t_j)$ and $\hat{Z}_k^i(t_j)$ take similar forms parallel to that of $\hat{X}_k^i(t_j)$ in accordance with the cyclic orders $X \to Y \to Z \to X$ and $x \to y \to z \to x$.

Theorem 2.2.3 *For the integral $J(F)$ with $\Delta_3^{2m}F \in C(K)$, there is a BTQF of the form (2.2.19) that possesses the homogeneous algebraic precision degree $3 \times \langle m-1\rangle + 1$ and uses $12n$ evaluation points on the edges of the cubic domain $[-1,1]^3$ as follows:*

$$
(\pm 1, \pm 1, t_j), (\pm 1, t_j, \pm 1), (t_j, \pm 1, \pm 1), \qquad j = 0, 1, \ldots, n-1,
$$

where t_j are the nodes of the integration formula (2.2.15).

Proof. It suffices to verify that (2.2.19) has the precision $3 \times \langle m-1 \rangle + 1$. Suppose that $F(x, y, z)$ is a polynomial of the form (2.2.14). Then (2.2.18) becomes exact since $\rho_m^2 = 0$. Moreover, it can be verified that $X_k(\pm 1, y, z)$, $Y_k(x, \pm 1, z)$, and $Z_k(x, y, \pm 1)$ are all of homogeneous degree $2 \times \langle m-1 \rangle + 1$. Hence, the right-hand side of (2.2.18) can be exactly evaluated by (2.2.16) in Lemma 2.2.2. This implies that (2.2.19) is exact for the polynomial F, and therefore proving the theorem. □

In particular, one may take (2.2.15) to be the Gauss–Legendre quadrature formula of the highest possible algebraic precision degree $m = 2n - 1$, so that n is the least number for a given m (an odd number). This leads to the following

Corollary 2.2.4 *Let $\Psi_{kij}(t_j)$ be defined by (2.2.20) with $m = 2n - 1$. Then we have a type of optimal BTQF of the form*

$$J(F) \approx \left(\frac{1}{(2n-1)!} \right) \sum_{k=0}^{5} \sum_{i=0}^{2n-2} \sum_{j=0}^{2n-2} \sum_{j=0}^{n-1} \Psi_{kij}(t_j) \tag{2.2.21}$$

that possesses the homogeneous precision $3 \times \langle 2n-2 \rangle + 1$ and uses the $12n$ edge points $(\pm 1, \pm 1, t_j)$, $(\pm 1, t_j, \pm 1)$, and $(t_j, \pm 1, \pm 1)$, $j = 0, 1, \ldots, n-1$, as evaluation points, where t_j are the zeros of the nth degree Legendre polynomial.

Evidently, the distribution of the evaluation points used in (2.2.21) is centrally symmetric with respect to the center of $[-1, 1]^3$, since the zeros of the Legendre polynomial are distributed symmetrically with respect to 0. Moreover, the number of evaluation points to be employed is generally much smaller than that used in the example at the end of Section 1.4 and in Example 2.1.2, although the formula is much more complicated in structure. In fact, for the given precision degree, $2m - 1$, of Example 2.1.2, $2m^2$ points are required, while for the corresponding precision, $3(2m-2) + 1$, of (2.2.21) (with $n = m$) only $12m$ points are needed. This also suggests that it may be better to use the latter quadrature formula instead of the former for $m > 6$.

Let us now return to the integral (2.2.1). Recall that the mapping $T : X \equiv (x, y, z) \to X \equiv (x, y, z)$ as mentioned in the beginning of this section is one-to-one, so that we may state the following.

Corollary 2.2.5 *Let ψ_i, ϕ_i $(i = 1, 2)$, and f be differentiable functions such that $\psi_i(x) \in C^{2m}[a, b]$, $\Delta_2^{2m} \phi_i(x, y) \in C([a, b] \times [\psi_1(x), \psi_2(x)])$, and $\Delta_3^{2m} f(x, y, z) \in C([a, b] \times [\psi_1(x), \psi_2(x)] \times [\phi_1(x, y), \phi_2(x, y)])$; and let $F(X) = T(f)(X) \cdot J(x, y)$. Then for integral (2.2.1) there is a kind of BTQF that may be derived from (2.2.21) in which $F(X)$ takes the place of $F(X^3)$. Finally, the $12n$ evaluation points on $\partial\Omega$ are determined through the inverse mapping T^{-1}.*

Moreover, with similar reasoning as that used in Theorem 2.1.2 and its proof, we can also get a kind of BTQF for (2.2.1) that may possess some preassigned degree of algebraic precision if ψ_i and ϕ_i $(i = 1, 2)$ are assumed to be polynomial functions. The detail is given by the following.

Theorem 2.2.6 *Suppose that ϕ_i and ψ_i $(i = 1, 2)$ are polynomial functions, and that $F(X) \equiv T(f)(X) \cdot J(x, y)$ is the transformed function of the integrand $f(X)$of (2.2.1) under the mapping T. Let there be given a formula of the form (2.2.21) in which $F(X)$ takes the place of $F(X^3)$. Denote r=max{ deg ψ_1, deg ψ_2} and s=max{ deg ϕ_1, deg ϕ_2}, and set*

$$t = \left[\frac{2n-2}{(1+r)(1+s)}\right] - 1, \tag{2.2.22}$$

where $[a]$ denotes the integral part of a. Then formula (2.2.21), considered as a formula in the (x, y, z)-system, is a BTQF for (2.2.1), with homogeneous precision of at least $3 \times \langle t\rangle$ with respect to $f(X)$.

Proof. The integration formula mentioned in Corollary 2.2.4 has the homogeneous precision degree $3 \times \langle 2n-2\rangle + 1$ with respect to the integrand function $F(X) \equiv T(f)(X) \cdot J(x, y)$. For convenience, we use $\overline{deg}(\cdot)$ to denote max{ $deg_x(\cdot), deg_y(\cdot), deg_z(\cdot)$}. Then, bearing in mind the meaning of r and s, we find

$$\begin{aligned}\overline{deg}F(X) &\leq \overline{deg}\,(T(f)(X)) + \overline{deg}J(x, y)\\ &\leq \left(\overline{deg}f\right)(1 + (1 + r)s) + (r + (1 + r)s)\\ &< \left(1 + \overline{deg}f\right)(1 + r)(1 + s).\end{aligned}$$

Thus if $\overline{deg}f \leq t$, then (2.2.22) implies that

$$\overline{deg}F(X) < (1 + t)(1 + r)(1 + s) \leq 2n - 2.$$

Consequently the integration formula in question becomes exact. In other words, it is exact for $deg_x f \leq t$, $deg_y f \leq t$, and $deg_z f \leq t$. Hence, at the least, it has the homogeneous precision $3 \times \langle t\rangle$ with respect to $f(X)$. The theorem is thereby proved. □

Certainly one can use the above results to construct various special boundary type formulas with given homogeneous precisions. In particular, one may get some centrally symmetric boundary type formulas if Ω is centrally symmetric.

Remark 2.2.1. The construction process used in this paper can be generalized to treat integrals with a weight function. For instance, if $V(X)$ is a given weight function, the required auxiliary function $G(X)$ should be chosen from the solution set of the equation $\Lambda G = V(X)$. Here one may take a differential operator of the form

$$\Lambda \equiv \partial^{|r|}/\partial x_1^{r_1} \cdots \partial x_n^{r_n}, \quad (|r| = r_1 + \cdots r_n)$$

if one wishes to construct a BTQF that has a precision just smaller than $\langle r_1, \dots, r_n\rangle$ but greater than $\langle r_1 - 1, \dots, r_n - 1\rangle$, etc. In the next section, we will use wavelets to construct the BTQF from a DRE for an integral with a weight function, $V(X) \in L^2$.

Remark 2.2.2. Instead of using the Gauss–Legendre formula in Example 2.1.2 and in Corollary 2.2.4, it is also possible to make use of some generalized *Gauss–Christoffel formula* or some variations of the Gaussian type formula by Stancu, Stroud, etc. (cf. [65], [66]), so that BTQFs with other forms may be generated. Of course, the degree of algebraic precision should be determined as required.

Remark 2.2.3. Generally, it is impossible to construct boundary type formulas of any preassigned precision using only the integrand function values at boundary points. For instance, such a formula for $J(F)$ of (2.2.4) given by [60] has algebraic precision degree 5 that cannot be exceeded in any way. In fact a simple counterexample given by $F(X) := (1-x^2)(1-y^2)(1-z^2)$ explains this point. However, if one does not insist on preserving a high degree of algebraic precision, it is still possible to get some "near-boundary type" formulas by just replacing all the partial derivatives with their corresponding numerical derivatives (of finite differences). Since such a process has to use numerical values of the integrand function at some interior points sufficiently close to the boundary, the formulas so obtained may therefore be called near-BTQFs. Such formulas should be applicable to some practical problems provided that enough information about the values of the related integrand function on or near the boundary is given.

2.3 Numerical integration associated with wavelet functions

In Remark 2.2.1, we discussed DRE and the corresponding BTQFs with the weight function $V(X)$. The idea is to solve $L(G)(X) = V(X)$ for $G(X)$ and to substitute it into DRE (1.2.3). First, we obtain a DRE for the integral $\int_\Omega FVdX$. To construct a BTQF of the integral, we need to construct a numerical quadrature formula for the boundary integral $\int_{\partial\Omega} W\, dS$ over the boundary of the domain Ω. A general process for the construction has been given in Sections 2.1 and 2.2. However, since the integral $\int_{\partial\Omega} W\, dS$ can be considered as an integral of a periodic function over $\partial\Omega$, we can use numerical techniques for an integral with a wavelet weight function to find a BTQF for $\int_\Omega FVdX$.

In this section, we will discuss two different methods of constructing numerical quadrature formulas using wavelet functions. The first method is based on the periodic wavelet expansion of weight functions. The second is based on the expansion of integrands (but not weight functions) using sampling functions. Here, a weight function is a bounded locally Lebesgue integrable function, which is not necessarily nonnegative.

A function $\psi(t) \in L^2(\mathbb{R})$ is called a *wavelet function* if it generates the basis $\psi_{mn}(t) = 2^{m/2}\psi(2^m t - n)$, $m, n \in Z$, of $L^2(\mathbb{R})$. A *multiresolution analysis*

(MRA) wavelet can be constructed from a *scaling function*, $\phi \in L^2(\mathbb{R})$, which generates an *MRA*.

We say ϕ is a scaling function, or equivalently, it generates an MRA, if it produces a nested sequence of closed subspaces $\{V_m\}_{m \in Z}$ of $L^2(\mathbb{R})$ such that

(i) $\cdots \subset V_{-1} \subset V_0 \subset V_1 \subset \cdots$,

(ii) $f(t) \in V_0 \leftrightarrow f(2t) \in V_1$,

(iii) $\overline{\cup_{m \in Z} V_m} = L^2(\mathbb{R})$,

(iv) $\cap_{m \in Z} V_m = \{0\}$, and

(v) $\{\phi(t-n)\}_{n \in Z}$ is a *Riesz basis* of V_0.

If condition (v) is changed to (v′), "$\{\phi(t-n)\}_{n \in Z}$ is an orthonormal basis of V_0," then the corresponding wavelet is called an *orthogonal MRA wavelet*. In this book, all wavelet functions will be MRA wavelets.

Denote $W_m = \overline{span\{\psi_{mn}(t)\}_{n \in Z}}$. The relationship between the scaling function ϕ and the corresponding wavelet ψ can be written as $V_{m+1} = V_m \oplus W_m$; i.e., the space W_m is the complement of V_m in V_{m+1}. ψ can be found from ϕ based on this relation. In particular, if ψ is an orthogonal (MRA) wavelet and its corresponding scaling function satisfies $\phi(t) = \sum_j c_j \phi(2t - j)$ (see condition (i) in the definition of MRA), then the *Fourier transform* of ψ can be described more directly as

$$\hat{\psi}(\xi) = e^{-i\xi/2} \overline{m_0\left(\frac{\xi}{2} + \pi\right)} \hat{\phi}\left(\frac{\xi}{2}\right), \tag{2.3.1}$$

where $m_0(\xi/2) = \hat{\phi}(\xi)/\hat{\phi}(\xi/2) = \sum_j \frac{c_j}{2} e^{-i\xi j/2}$.

Obviously, that the scaling function set $\{\phi(t-n)\}_{n \in Z}$ is orthonormal and is equivalent to that of the following identity holds for $a.e.\xi \in \mathbb{R}$.

$$\sum_{j=-\infty}^{\infty} |\hat{\phi}(\xi + 2\pi j)|^2 = 1. \tag{2.3.2}$$

In fact, equation (2.3.2) can be obtained by using the *Parseval identity* $(f, g) = (\hat{f}, \hat{g})/2\pi$. Substituting the relation $\hat{\phi}(\xi) = m_0(\xi/2)\hat{\phi}(\xi/2)$ into equation (2.3.2) and separating the summation into two parts, one for odd j and another for even j, we have

$$|m_0(\xi)|^2 + |m_0(\xi + \pi)|^2 = 1. \tag{2.3.3}$$

From the above equation and the two-scale relation

$$\hat{\phi}(\xi) = m_0(\xi/2)\hat{\phi}(\xi/2),$$

we immediately know $\hat{\phi}(0) = 1$ and $\hat{\phi}(\pi) = 0$.

In the following, we will give some well-known scaling functions and their associated wavelets.

Example 2.3.1. The *Shannon scaling function* $\phi(t) = \sin(\pi t)/(\pi t)$ has the corresponding wavelet

$$\psi(t) = \frac{\sin \pi \left(t - \frac{1}{2}\right) - \sin 2\pi \left(t - \frac{1}{2}\right)}{\pi \left(t - \frac{1}{2}\right)}.$$

Example 2.3.2. If a scaling function is defined by

$$\hat{\phi}(\xi) = \left(\int_{\xi-\pi}^{\xi+\pi} h(u)du\right)^{1/2},$$

where h is a symmetric and nonnegative integrable function that satisfies (i) $\text{supp}(h) \in [-\pi/3, \pi/3]$ and (ii) $(h, 1) = 1$, then the corresponding wavelet is called *the Meyer wavelet.* Obviously, $\hat{\phi}(\xi) = 1$ for $|\xi| \le 2\pi/3$ and $\text{supp}\hat{\phi}(\xi) = [-4\pi/3, 4\pi/3]$.

Example 2.3.3. The *Haar scaling function* $\phi(t) = \chi_{[0,1]}(t)$, the characteristic function of the interval $[0, 1]$, is associated with the *Haar wavelet* $\psi(t) = \phi(2t) - \phi(2t - 1)$.

Example 2.3.4. The *B-spline* of order m is defined by

$$N_m(t) = (N_{m-1} * N_1)(t) = \int_0^1 N_{m-1}(x - t)dt, \qquad m \ge 2, \tag{2.3.4}$$

where $N_1(t) = \chi_{[0,1)}(t)$ is the characteristic function of the interval $[0, 1)$. It is obvious that

$$\hat{N}_m(\xi) = \left(\hat{N}_1(\xi)\right)^m = \left(\frac{1 - e^{-i\xi}}{i\xi}\right)^m. \tag{2.3.5}$$

Although N_m is a scaling function, it does not generate an orthogonal MRA. However, condition (2.3.2) tells us that an orthogonal scaling function associated with N_m can be defined as

$$\hat{\phi}_m(\xi) = \frac{\hat{N}_m(\xi)}{\sum_{j=-\infty}^{\infty} |\hat{N}_m(\xi + 2\pi j)|^2}. \tag{2.3.6}$$

From [9],

$$\sum_{j=-\infty}^{\infty} |\hat{N}_m(2\xi + 2\pi j)|^2 = \frac{-\sin^{2m} \xi}{(2m - 1)!} \frac{d^{2m-1}}{d\xi^{2m-1}} \cot \xi. \tag{2.3.7}$$

Therefore, for $m = 1$ and $m = 2$, we have $\sum_{j=-\infty}^{\infty} |\hat{N}_1(\xi + 2\pi j)|^2 = 1$ and

$$\sum_{j=-\infty}^{\infty} |\hat{N}_2(\xi + 2\pi j)|^2 = 1 - \frac{2}{3} \sin^2 \left(\frac{\xi}{2}\right), \tag{2.3.8}$$

respectively. It follows that $\phi_1(t) = N_1(t)$. And from equations (2.3.5), (2.3.6), and (2.3.8) we may define the scaling function $\phi_2(t)$ as

$$\hat{\phi}_2(\xi) = \frac{\sin^2(\xi/2)}{(\xi/2)^2}\left(1 - \frac{2}{3}\sin^2(\xi/2)\right)^{-1/2}. \tag{2.3.9}$$

Example 2.3.5. We can also consider the equation $\phi(t) = \sum_j c_j\phi(2t-j)$ as the recurrence relation

$$\phi^n(t) = \sum_{j\in J} c_j\phi^{n-1}(2t-j) \tag{2.3.10}$$

with the initial function $\phi^0(t)$. As an example, for $\chi_{[0,1)}(t)$ and any finite set J, the *Daubechies wavelets*, ψ, can be constructed based on this process if the coefficient set $\{c_j\}_{j\in J}$ are solvable. To find the coefficient set, we need to solve a system that consists of the equations $\hat{\phi}(0) = 1$, $\hat{\phi}(\pi) = 0$, $(\phi(t), \phi(t-n)) = \delta_{0n}$, $n \in Z$, and the vanishing moment conditions $(t^k, \psi) = 0$ for some $k \in Z_+$. The first two equations come from condition (2.3.3). The third equation is from the orthogonality of $\{\phi(t-n)\}_{n\in Z}$. For instance, if we set $k = 1$ and $J = \{0, 1, 2, 3\}$, then the above system implies

$$\sum_{j\in J} c_j = 2,\ \sum_{j\in J}(-1)^j c_j = 0,\ \sum_{j\in J} c_j^2 = 2,\ \sum_{j\in J}(-1)^j jc_j = 0.$$

The corresponding solutions are $c_0 = (1+\sqrt{3})/4$, $c_1 = (3+\sqrt{3})/4$, $c_2 = (3-\sqrt{3})/4$, and $c_3 = (1-\sqrt{3})/4$.

We now give the first method for constructing numerical quadrature formulas of the integral $\int_{\partial\Omega} W\,dS$ shown in DRE (1.2.3) with wavelet functions. Here, we assume that $\partial\Omega$ is a piecewise C^1 continuous surface or a piecewise C^1 continuous simple closed curve. This construction can be immediately generalized to functions in higher dimensions. Therefore, without a loss of generality, we may assume that $\partial\Omega$ is a piecewise C^1 continuous simple closed curve defined by $c(t) : [0, 1] \to \mathbb{R}$, $c(t) = (x(t), y(t))$, $c(0) = c(1)$. In addition, $F(x, y) = F(x(t), y(t))$ and $G(x, y) = G(x(t), y(t))$ are functions with period 1 in the space C^{m-1} – periodic (the subspace of periodic functions in the space C^{m-1}). Hence $\int_{\partial\Omega} W\,dS$ can be written as

$$\int_{\partial\Omega} W\,dS = \sum_{k=0}^{m_1-1}(-1)^k\int_0^1 D^{(k,0)}F(x(t), y(t))D^{(m_1-k-1,m_2)}$$
$$G(x(t), y(t))y'(t)\,dt + \sum_{k=0}^{m_2-1}(-1)^{m_1+k+1}\int_0^1 D^{(m_1,k)}$$
$$F(x(t), y(t))D^{(0,m_2-k-1)}G(x(t), y(t))x'(t)\,dt. \tag{2.3.11}$$

The integrals in (2.3.11) can be considered as integrals of periodic weight functions of period 1: $D^{(m_1-k-1,m_2)} \cdot G(x(t), y(t))y'(t)$, $k = 0, 1, \ldots, m_1 - 1$; and $D^{(0,m_2-k-1)}G(x(t), y(t))x'(t)$, $k = 0, 1, \ldots, m_2 - 1$. All these weight functions can be expanded as wavelet series in terms of periodic wavelets of period 1. Therefore, the integrals in (2.3.11) can be decomposed to integrals of the form $\int_0^1 f(t)\psi^*(t)\,dt$, $\psi^*(t)$ being a periodic wavelet of period 1 and $f(t)$ being a periodic function of period 1. Thus we focus on two problems: (a) approximating continuous functions with periodic wavelet series, and (b) numerical quadrature formulas for the integral $\int_0^1 f(t)\psi^*(t)\,dt$.

To approximate continuous functions with periodic wavelet series, we begin with the collection, S_r, of rapidly decreasing C^r continuous functions defined on $\mathbb{R}$; i.e, functions that satisfy

$$|g^{(k)}(t)| \le C_{\ell k}(1 + |t|)^{-\ell}, \quad k = 0, 1, \ldots, r, \quad \ell > 0, \quad t \in \mathbb{R}. \tag{2.3.12}$$

All scaling functions shown in Examples 2.3.1–2.3.5 are in S_r for some r. We also denote by $S_{0,p}$ $(p > 1)$ the space of all functions satisfying

$$|g(t)| \le C_{pk}(1 + |t|)^{-p}, \qquad t \in \mathbb{R}.$$

Let $\phi \in S_r$ be a scaling function associated with multiresolution analysis wavelet ψ. We define ϕ^*_{mn} and ψ^*_{mn} to be the periodized versions of $\phi_{mn}(t) = 2^{m/2}\phi(2^m t - n)$ and $\psi_{mn}(t) = 2^{m/2}\psi(2^m t - n)$, namely

$$\phi^*_{mn}(t) := \sum_k \phi_{mn}(t - k), \quad \psi^*_{mn} := \sum_k \psi_{mn}(t - k). \tag{2.3.13}$$

Both ϕ^*_{mn} and ψ^*_{mn} are periodic functions of period 1. Obviously $\phi^*_{00}(t) = 1$ and $\phi^*_{mn}(t) \in \text{span}\,\{\phi^*_{0k}(t)\}$ when $m < 0$. Thus, we only need to consider $\phi^*_{mn}(t)$ for $m \ge 0$. We denote by V^*_m the space spanned by $\{\phi^*_{mn}(t)\}_{n \in Z}$. V^*_m is clearly a finite dimensional space since $\phi^*_{mn} = \phi^*_{m,n+2^m}$. In fact, $\phi^*_{m,0}, \phi^*_{m,1}, \ldots, \phi^*_{m,2^m-1}$ are an orthogonal basis of V^*_m; the orthogonality is given by $\int_0^1 \phi^*_{m,0}\phi^*_{m,n} = 0$, $0 < n < 2^m$. These V^*_m's satisfy the nested property $V^*_0 \subset V^*_1 \subset \cdots \subset V^*_m \subset \cdots$ and $\overline{\bigcup_m V^*_m} = L^2(0, 1)$.

In [64], $\{V^*_m\}_{j=0}^\infty$ is called a *periodic multiresolution analysis (PMRA)*. Some approximation properties of PMRA are given in [69] and [64]. For example, it can be shown that the V^*_ms are reproducing kernel Hilbert spaces with the reproducing kernels

$$q^*_m(x, t) = \sum_{n=0}^{2^m-1} \phi^*_{mn}(x)\phi^*_{mn}(t),$$

where $x, t \in [0, 1]$. In addition, it can be shown that

$$\{q_m^*(x, t) = \chi_{[0,1)}(x)\chi_{[0,1)}(t)\}$$

is a quasi-positive delta sequence. Hence we have the following result.

Proposition 2.3.1 *Let $g(t)$ be continuous and periodic, then the projection, g_m^*, of g onto V_m^* converges to g uniformly. Here, V_m^*, $m \in Z$, is generated by $\phi^* \in S_{0,p}$, $p > 1$.*

Proof. From Corollary 3.18 in [33], we immediately have Proposition 2.3.1. A similar result for $\phi^* \in S_r$ was given by Proposition 7.7 in [69]. □

The quantitative estimate of the approximation shown in Proposition 2.3.1 can be found by means of the following theorem, from [64].

Theorem 2.3.2 *If $\phi^* \in S_{0,q}$, $q > 1$, $\psi^* \in C^m(\mathbb{R})$ with $(\psi^*)^{(\ell)}$ bounded for $\ell \leq m$, and $\psi^* \in S_{0,r}$, $r > m + 1$, then for any real number $p \in [0, \infty]$, any positive integer k with $k \leq m+1$ and $k < r - 1$, and all functions $f \in L_p([0, 1])$ ($f \in C([0, 1])$ for $p = \infty$), we have*

$$\|f - S_N(f)\|_{L_p} \leq C(p, r, m)\omega_k\left(f, \frac{1}{N}\right)_p, \quad N = 1, 2, \dots,$$

where

$$S_N(f) = \sum_{n=0}^{2^m-1} \langle f, \phi_{mn}^* \rangle \phi_{mn}^* + \sum_{n=0}^{L} \langle f, \psi_{mn}^* \rangle \psi_{mn}^*,$$

$N = 2^m + L$, $0 \leq L < 2^m - 1$, and

$$\omega_k(f, h)_p = \sup_{|t| \leq h} \|\Delta_t^k f\|_{L_p}$$

is the kth modulus of smoothness of f.

Remark 2.3.1. The corresponding wavelets themselves are simpler in the periodic case as well. We denote $\psi_0(t) = \phi_{00}^*(t) = 1$ and $\psi_{2^m+k}(t) = \psi_{m,k}^*(t)$, $0 \leq m$, $k = 0, 1, \dots, 2^m - 1$. The set $\{\psi_n\}$ is clearly an orthonormal basis of $L^2(0, 1)$ consisting of periodic functions. In particular, the set $\{\psi_0, \psi_1, \dots, \psi_{2^m-1}\}$ is also an orthonormal basis of V_m^* since there are 2^m of them and they are orthonormal. The orthogonal expansion of $g \in L^2(0, 1)$ is given by $\sum_{n=0}^{\infty} \langle g, \psi_n \rangle \psi_n$. If g is continuous, from the proposition, we have $g = \sum_{n=0}^{\infty} \langle g, \psi_n \rangle \psi_n$.

Remark 2.3.2. The periodic wavelet series expansion can be extended to the space $H^{m-1}(0, 1)$ – periodic, the subspace of periodic functions of period 1 in the Sobolev space H^{m-1} (see Jaffard and Laurençot [48] and Meyer [54]). The corresponding decomposition formulas can also be found in [48] and [54].

Remark 2.3.3. Obviously, the property of uniform convergence does not hold for trigonometric Fourier series expansions.

We now discuss the numerical quadrature formula of the integral $\int_0^1 f(t) \psi^*(t)dt$. Here $f(t)$ and $\psi^*(t)$ are, respectively, a periodic function and a periodic wavelet of period 1. We first give a general quadrature for the integral $\int_{-\infty}^{\infty} f(t)\psi(t)\,dt$ with the nonperiodic wavelet weight function $\psi(t)$.

Theorem 2.3.3 *Let $\psi \in S_r$, with $\psi_{mn}(t) = 2^{m/2}\psi(2^m t - n)$ being an orthonormal system in $L^2(\mathbb{R})$, and $\int_{-\infty}^{\infty} t^{r+1}\psi(t)\,dt = A \neq 0$. Then for any set of distinct evaluation points $\{t_i, i = 0, 1, \ldots, r+1\}$ and $f \in C^{r+2}$,*

$$\int_{-\infty}^{\infty} f(t)\psi(t)\,dt \approx \sum_{i=0}^{r+1} c_i f(t_i), \tag{2.3.14}$$

where

$$c_i = A \prod_{j=0, j\neq i}^{r+1} \frac{1}{t_i - t_j}. \tag{2.3.15}$$

In addition, if $\int_{-\infty}^{\infty} x^{r+2}\psi(t)\,dt = B$ and $|f^{(r+2)}(t)| \leq M$, then the error in quadrature formula (2.1.3) is

$$\left| \int_{-\infty}^{\infty} f(t)\psi(t)\,dt - \sum_{i=0}^{r+1} c_i f(t_i) \right| \leq \frac{M}{(r+2)!}\left[|B| + |A| \left| \sum_{j=0}^{r+1} t_j \right| \right].$$

Proof. For the given set of distinct evaluation points, we have the Lagrange interpolation

$$f(t) \approx \sum_{i=0}^{r+1} f(t_i) l_i(t),$$

where

$$l_i(t) = \prod_{j=0, j\neq i}^{r+1} \frac{t - t_j}{t_i - t_j}.$$

The error in the interpolation is

$$\frac{1}{(r+2)!} \prod_{j=0}^{r+1} (t - t_j) f^{(r+2)}(\zeta)$$

for some ζ. Thus

$$\int_{-\infty}^{\infty} t^k \psi(t)\, dt \approx \sum_{i=0}^{r+1} f(t_i) \int_{-\infty}^{\infty} l_i \psi(t)\, dt.$$

Since $\psi(t) \in S_r$ and the $\{\psi_{m,n}\}$ is an orthonormal system in $L^2(\mathbb{R})$, we have (see Theorem 3.1 in [69])

$$\int_{-\infty}^{\infty} t^k \psi(t)\, dt = 0, \quad k = 0, 1, \dots, r.$$

Therefore,

$$\int_{-\infty}^{\infty} l_i(t)\psi(t)\, dt = \int_{-\infty}^{\infty} \left(\prod_{j=0}^{r+1} \frac{1}{t_i - t_j} \right) t^{r+1} \psi(t)\, dt = A \prod_{j=0, j\neq i}^{r+1} \frac{1}{t_i - t_j}.$$

□

Remark 2.3.4. If $\int_{-\infty}^{\infty} t^k \psi(t)\, dt = 0$ for $k = 0, 1, \dots, r, \dots, m$, but $\int_{-\infty}^{\infty} x^{m+1} \psi(x)\, dx \neq 0$, then we can increase the number of evaluation points to $m+2$ and construct the corresponding quadrature formula similarly.

Remark 2.3.5. If $\psi_{mn}(t)$ is not an orthonormal basis in $L^2(\mathbb{R})$, then we can use the regularity of the duals of $\psi_{mn}(t)$ to construct the numerical quadrature of $\int_{-\infty}^{\infty} f(t)\psi(t)\, dt$ (see Theorem 5.5.1 in Daubechies [11]).

Corollary 2.3.4 *Let $\psi \in S_r$, with $\psi_{mn}(t) = 2^{m/2}\psi(2^m t - n)$ being an orthonormal system in $L^2(\mathbb{R})$, and $\int_{-\infty}^{\infty} t^{r+1}\psi(t)\, dt = A \neq 0$. Also let $\psi^*(t) = \sum_k \psi(t-k)$ be the periodized version of ψ with period 1. Then for a C^{r+2} continuous periodic function f of period 1, we have the quadrature formula*

$$\int_0^1 f(t)\psi^*(t)\, dt \approx \sum_{i=0}^{r+1} c_i f(t_i), \tag{2.3.16}$$

where

$$c_i = A \prod_{j=0, j\neq i}^{r+1} \frac{1}{t_i - t_j} \tag{2.3.17}$$

and $\{t_i : i = 0, 1, \dots, r+1\}$ are $r+2$ distinct real numbers.

Proof. First, we note that, if $t_i \notin [0, 1)$, we can always find $t_i' \in [0, 1)$ such that $f(t_i) = f(t_i')$ because of the periodicity of f. Second, we have

$$\begin{aligned} \int_0^1 f(t)\psi^*(t)\,dt &= \int_0^1 f(t)\sum_k \psi(t-k)\,dt \\ &= \sum_k \int_k^{k+1} f(t+k)\psi(t)\,dt = \int_{-\infty}^{\infty} f(t)\psi(t)\,dt. \end{aligned}$$

Hence, (2.3.16) and (2.3.17) hold because of (2.3.14) and (2.3.15) respectively. □

We now go back to the integrals in (2.3.11). The weight function in each integral can be expanded in terms of sufficiently smooth periodic wavelet functions, such as the C^r continuous periodic wavelet $\psi^* = \sum_k \psi(t-k)$, $\psi \in S_r$. Then, for each integral with the form $\int_0^1 f(t)\psi_{mn}^*(t)\,dt$ in (2.3.11), we make use of quadrature (2.3.16), and the BTQF for $\int_\Omega F\,dX$ is eventually obtained by replacing the integral $\int_{\partial\Omega} W\,dS$ in (1.4.2) with the quadrature constructed through the above process.

The second method for constructing numerical quadrature formulas is based on an extension of the Shannon sampling theorem in V_m, the L^2 closure of $\text{span}\{\phi_{mn}(t)\}_{n\in Z}$. Here, ϕ is a scaling function.

If $f(t)$ is in $C \cap L^2(\mathbb{R})$, then the *Shannon sampling theorem* tells us that $f(t)$ can be recovered from its values at jT, $j \in Z$, $T = \pi/\sigma$, $\sigma > 0$,

$$f(t) = \sum_{j=-\infty}^{\infty} f(jT)\frac{\sin\sigma(t-jT)}{\sigma(t-jT)}. \tag{2.3.18}$$

If $\sigma = 2^m\pi$, noting that the Shannon scaling function is $\phi(t) = \frac{\sin \pi t}{\pi t}$ (see Example 2.3.1), then equation (2.3.18) becomes

$$f(t) = \sum_{j=-\infty}^{\infty} f\left(\frac{j}{2^m}\right)\phi(2^m t - j).$$

The above equation can be considered as the Shannon sampling theorem in V_m. There is no doubt that a sampling theorem is a source for constructing a numerical quadrature formula. Thus, a question arises: can the Shannon sampling theorem be extended to other scaling functions under some broad conditions (so that we may use the theorem to construct different quadrature formulas)? The answer is yes and the details are given in the following theorem.

Theorem 2.3.5 *[69] Let $\phi(t)$ be a scaling function in S_r, $r > 1$, such that*

$$\hat{\phi}^*(\xi) := \sum_j \phi(j)e^{-i\xi j} \neq 0, \tag{2.3.19}$$

for all $\xi \in \mathbb{R}$. Then for any $f \in V_0$,

$$f(t) = \sum_j f(j)\bar{\phi}(t-j), \qquad t \in \mathbb{R}. \tag{2.3.20}$$

The convergence is uniform on $\mathbb{R}$ and the sampling function $\bar{\phi}$ is defined by

$$\hat{\bar{\phi}}(\xi) = \hat{\phi}(\xi)/\hat{\phi}^*(\xi), \qquad \xi \in \mathbb{R}. \tag{2.3.21}$$

Proof. Obviously, $q(x,t) = \sum_j \phi(x-j)\phi(t-j)$ is the reproducing kernel of V_0; i.e., $(f(t), q(x,t)) = f(x)$. It can also be shown that there exists a basis, denoted by $\{\bar{\phi}(t-j)\}_{j\in Z}$, in V_0 that is biorthogonal to $q(t,n)$. Hence,

$$f(t) = \sum_j (f(t), q(t,j))\bar{\phi}(t-j) = \sum_j f(j)\bar{\phi}(t-j).$$

To establish equation (2.3.21), we substitute $\phi(t)$ for $f(t)$ and perform a Fourier transform on the above equation, giving

$$\hat{\phi}(\xi) = \sum_j \phi(j)e^{-i\xi j}\hat{\bar{\phi}}(\xi) = \hat{\phi}^*(\xi)\hat{\bar{\phi}}(\xi).$$

We now show that the convergence in expression (2.3.20) is uniform. First, $q(t,t)\int |q(t,x)|^2dx = \sum_j \phi^2(t-j)$ is periodic and is therefore uniformly bounded. Noting $(f(t), q(x,t)) = f(x)$ and using the Cauchy–Schwartz inequality, we have

$$|f(x) - \sum_{j=-N}^{N} f(j)\bar{\phi}(x-j)| \le (q(t,t))^{1/2}\, \|f(x) - \sum_{j=-N}^{N} f(j)\bar{\phi}(x-j)\|_2.$$

This completes the proof. □

According to [69], there are three ways to construct $\hat{\phi}^*(\xi)$ from a certain $\phi(t)$. In turn, the sampling function $\bar{\phi}(t)$ can be found by using formula (2.3.21). The first method utilizes the definition of $\hat{\phi}^*(\xi)$ shown in (2.3.19). Consider the Daubechies wavelet in Example 2.3.5. In equation (2.3.10), substituting $t = 0, 1, 2$, and 3 in turn (while noting that $c_0 = (1+\sqrt{3})/4$, $c_1 = (3+\sqrt{3})/4$, $c_2 = (3-\sqrt{3})/4$, and $c_3 = (1-\sqrt{3})/4$), we obtain a system that can be solved for $\phi(t)$, $t = 0, 1, 2, 3$. Hence,

$$\hat{\phi}^*(\xi) = \frac{1}{2\sqrt{3}}\left((\sqrt{3}-1)e^{-i\xi} + (\sqrt{3}+1)e^{-2i\xi}\right)$$

and

$$\hat{\phi}^{*-1}(\xi) = \frac{2\sqrt{3}}{1+\sqrt{3}}\sum_{j=0}^{\infty}\left(\frac{1-\sqrt{3}}{1+\sqrt{3}}\right)^j e^{-i(j-1)\xi}.$$

It follows that the corresponding *Daubechies sampling function* is

$$\bar{\phi}(t) = \frac{2\sqrt{3}}{1+\sqrt{3}} \sum_{j=0}^{\infty} \left(\frac{1-\sqrt{3}}{1+\sqrt{3}} \right)^j \phi(t+j+2).$$

Similarly, for the Haar scaling function $\phi(t)$, $\bar{\phi}(t) = \phi(t)$.

The second method for constructing $\bar{\phi}(t)$ is based on the formula

$$\hat{\phi}^*(\xi) = \sum_{j=-\infty}^{\infty} \hat{\phi}(\xi + 2\pi j), \tag{2.3.22}$$

where $\phi \in S_r, r \geq 2$. Clearly, the series on the right-hand side of equation (2.3.22) is uniformly convergent. In addition, the Fourier coefficients of the function defined by the series on the right-hand side of (2.3.22) are $\phi(-j)$, $j \in Z$. Hence the corresponding Fourier series expansion of the function is $\sum_j \phi(-j)e^{i\xi j} = \sum_j \phi(j)e^{-i\xi j}$, which is equal to $\hat{\phi}^*(\xi)$ from (2.3.19).

As an example, let us consider the B-spline scaling function $\phi_2(t)$ defined in (2.3.9). We have $\hat{\phi}_2^*(\xi) = \sum_j \hat{\phi}_2(\xi + 2\pi j) = \left(1 - \frac{2}{3}\sin^2\frac{\xi}{2}\right)^{-1/2}$ and $\hat{\bar{\phi}}_2(\xi) = \hat{\phi}_2(\xi)/\hat{\phi}_2^*(\xi) = \left(\frac{\sin(\xi/2)}{\xi/2}\right)^2$. Therefore, the *B-spline sampling function* is $\bar{\phi}_2(t) = (1 - |t|)\chi_{(-1,1)}(t)$.

The third method for constructing a sampling function is shown as follows for the Meyer scaling function. From the definition of the Meyer scaling function given in Example 2.3.2, we have

$$\hat{\phi}^*(\xi) = \hat{\phi}(\xi) + \hat{\phi}(\xi - 2\pi) + \hat{\phi}(\xi + 2\pi),$$

for all $|\xi| \leq 4\pi/3$. Hence, the Fourier transform of the *Meyer sampling function* is

$$\hat{\bar{\phi}}(\xi) = \frac{\hat{\phi}(\xi)}{\hat{\phi}(\xi) + +\hat{\phi}(\xi - 2\pi) + \hat{\phi}(\xi + 2\pi)}, \qquad \xi \in \mathbb{R}.$$

$\hat{\bar{\phi}}(\xi) = 1$ when $|\xi| \leq 2\pi/3$.

We now construct numerical quadrature formulas by using sampling functions. First, we give a formula for the integrals of periodic functions with period 1.

Theorem 2.3.6 *Let $g(t) \in L^1([0,1])$, and let $f \in C([0,1])$ be a periodic function with period 1. Also denote by $f_m^* = P_m f$ the projection of f on the space $V_m^* = span\{\bar{\phi}_{m,i}^*, 0 \leq i \leq 2^m - 1\}$; $\bar{\phi}_{m,i}^* \in S_{0,p}$, $p > 1$, is defined similarly as in (2.3.13) with only the replacement of $\phi = \bar{\phi}$. Then the quadrature formula*

$$\int_0^1 g(t)f(t)dt \approx \frac{1}{2^m} \sum_{j=0}^{2^m-1} a_{m,j} f\left(\frac{j}{2^m}\right) \tag{2.3.23}$$

is convergent when $m \to \infty$ *and is exact for all functions in* V_m^*. *Here,*

$$a_{m,j} = 2^m \int_0^1 g(t)\bar{\phi}^*(2^m t - j)dt.$$

Proof. From Proposition 2.3.1, we have $\|f_m - f\|_\infty \to 0$ as $m \to \infty$, where f_m is the projection of f on the space V_m, which is generated by a scaling function $\phi \in S_{0,p}$, $p > 1$. Thus, for any given $\delta > 0$, there exists an integer $M > 0$ such that

$$|f(t) - f_m(t)| < \delta,$$

for all $m > M$ and $t \in [0, 1]$.

If $f \in V_m^*$, then

$$f(t) = \sum_{j=0}^{2^m-1} f\left(\frac{j}{2^m}\right)\bar{\phi}^*(2^m t - j).$$

Multiplying by $g(t)$ and then taking the integral from 0 to 1 on both sides of the above equation, we find quadrature formula (2.3.23) to hold exactly for $f = f_m$. Therefore, for $m > M$,

$$\begin{aligned}
&\left|\int_0^1 g(t)f(t)dt - \frac{1}{2^m}\sum_{j=0}^{2^m-1} a_{m,j} f\left(\frac{j}{2^m}\right)\right| \\
\le\ & \left|\int_0^1 g(t)\left(f(t) - f_m(t)\right)dt\right| \\
&+ \left|\frac{1}{2^m}\sum_{j=0}^{2^m-1} a_{m,j} f_m\left(\frac{j}{2^m}\right) - \frac{1}{2^m}\sum_{j=0}^{2^m-1} a_{m,j} f\left(\frac{j}{2^m}\right)\right| \\
\le\ & \|g\|_{L^1}\delta + \frac{1}{2^m}\sum_{j=0}^{2^m-1}\left|a_{m,j}\left(f_m\left(\frac{j}{2^m}\right) - f\left(\frac{j}{2^m}\right)\right)\right| \\
\le\ & \|g\|_{L^1}\delta + \left(\frac{1}{2^m}\sum_{j=0}^{2^m-1}|a_{m,j}|\right)\delta.
\end{aligned}$$

$\sum_{j=0}^{2^m-1}|\bar{\phi}^*(t-j)|$ is bounded because it is a continuous periodic function with the period 1. Hence, we have

$$C = \sup_{t\in\mathbb{R}}\left\{\sum_{j=0}^{2^m-1}|\bar{\phi}^*(t-j)|\right\} < \infty.$$

With the constant C as defined above, we obtain

$$\begin{aligned}\sum_{j=0}^{2^m-1}|a_{m,j}| &\le \sum_{j=0}^{2^m-1}\int_0^{2^m} g(2^{-m}t)|\bar{\phi}^*(t-j)|dt \\ &= \int_0^{2^m} g(2^{-m}t)\left(\sum_{j=0}^{2^m-1}|\bar{\phi}^*(t-j)|\right)dt \\ &\le C2^m\|g\|_{L^1}.\end{aligned}$$

Therefore,

$$\begin{aligned}&\left|\int_0^1 g(t)f(t)dt - \frac{1}{2^m}\sum_{j=0}^{2^m-1} a_{m,j}f\left(\frac{j}{2^m}\right)\right| \\ \le\ & \|g\|_{L^1}\delta + \left(\frac{1}{2^m}\sum_{j=0}^{2^m-1}|a_{m,j}|\right)\delta \\ \le\ & (1+C)\|g\|_{L^1}\delta.\end{aligned}$$

Setting $\delta = \epsilon/(1+C)\|g\|_{L^1}$ gives

$$\left|\int_0^1 g(t)f(t)dt - \frac{1}{2^m}\sum_{j=0}^{2^m-1} a_{m,j}f\left(\frac{j}{2^m}\right)\right| \le \epsilon$$

for $m > M$ and completing the proof of the theorem. □

By applying Theorem 2.3.6 to the integrand of each boundary integral in (2.3.11), we thus obtain a BTQF.

At the end of this section, we will extend Theorem 2.3.6 to integrals over $\mathbb{R}$ for fast decay functions f; i.e., we will construct a quadrature formula for the integral $\int_{-\infty}^{\infty} g(t)f(t)dt$ using the sampling function $\bar{\phi}$. Thus, we have the following result.

Theorem 2.3.7 *Let $g(t) \in L^1(\mathbb{R})$. If $f \in C(\mathbb{R})$, and $|f(t)| \le C/(1+|t|^r)$ $(r > 2)$, then the quadrature formula*

$$\int_{-\infty}^{\infty} g(t)f(t)dt \approx \frac{1}{2^m}\sum_{j=-\infty}^{\infty} a_{m,j}f\left(\frac{j}{2^m}\right) \tag{2.3.24}$$

with

$$a_{m,j} = 2^m\int_{-\infty}^{\infty} g(t)\bar{\phi}(2^m t - j)dt \tag{2.3.25}$$

is convergent as $m \to \infty$ and is exact for all functions in V_m, which is the space generated by a sampling function $\bar{\phi} \in S_r$, $r > 1$.

Proof. The proof of Theorem 2.3.7 is similar to that of Theorem 2.3.6. First, we have $f_m \to f$ as $m \to \infty$ uniformly on any compact subset of $\mathbb{R}$. Noting the decay of f, the convergence is uniform on the entire $\mathbb{R}$. Hence, for any $\delta > 0$, there exists M such that $|f_m(t) - f(t)| < \delta$ for all $t \in \mathbb{R}$ and $m > M$. Here, f_m is the projection of f on V_m. Secondly, from $V_m = span\ \{\bar{\phi}_{mn} : n \in Z\}$ and $\bar{\phi} \in S_r$, we have that

$$f_m(t) = \sum_{j=-\infty}^{\infty} f_m\left(\frac{j}{2^m}\right)\bar{\phi}(2^m t - j)$$

is uniformly convergent on $\mathbb{R}$ and is bounded by $1/(1+|t|^r)$. Using the Lebesgue dominated convergence theorem yields

$$\int_{-\infty}^{\infty} g(t) f_m(t) = \frac{1}{2^m}\sum_{j=-\infty}^{\infty} a_{m,j} f_m\left(\frac{j}{2^m}\right).$$

Next, similar to the argument in the proof of Theorem 2.3.6, we obtain

$$\begin{aligned}&\left|\int_{-\infty}^{\infty} g(t) f(t)dt - \frac{1}{2^m}\sum_{j=-\infty}^{\infty} a_{m,j} f\left(\frac{j}{2^m}\right)\right| \\ \leq\ & (1+C)\|g\|_{L^1}|f(t) - f_m(t)| \leq (1+C)\|g\|_{L^1}\delta \end{aligned} \tag{2.3.26}$$

for $m > M$. Here, C is a constant defined by

$$C = \sup_{t\in\mathbb{R}}\left\{\sum_{j=-\infty}^{\infty} |\bar{\phi}(t-j)|\right\}.$$

□

Theorem 2.3.6 was given in [61] for the Meyer sampling function, and the error bound of the corresponding quadrature formula was described as follows (see [61]).

Theorem 2.3.8 *If $f \in C \cap L^2(\mathbb{R})$ and its Fourier transform satisfies*

$$|\hat{f}(\xi)| \leq Ce^{-\lambda|\xi|}, \quad \lambda > 0,$$

then for the quadrature formula (2.3.24) constructed by using the Meyer sampling function, the estimate is

$$\left|\int_{-\infty}^{\infty} g(t) f(t)dt - \frac{1}{2^m}\sum_{j=-\infty}^{\infty} a_{m,j} f\left(\frac{j}{2^m}\right)\right| \leq Ke^{-2^{m+1}\lambda\pi/3}. \tag{2.3.27}$$

Proof. From Lemma 8.7 in [69],

$$\|f - f_m\|_\infty = O\left(e^{-2^{m+1}\lambda\pi/3}\right), \tag{2.3.28}$$

where f_m is the projection of f on V_m, which is generated by the Meyer sampling function $\bar{\phi}$. The function f satisfies the conditions of Theorem 2.3.7. Hence, we have quadrature formula (2.3.24) and its error estimate,

$$\begin{aligned}&\left|\int_{-\infty}^{\infty} g(t)f(t)dt - \frac{1}{2^m}\sum_{j=-\infty}^{\infty} a_{m,j} f\left(\frac{j}{2^m}\right)\right|\\ \le\ & (1+C)\|g\|_{L^1}\|f(t) - f_m(t)\|_\infty.\end{aligned}$$

By substituting equation (2.3.28) into the above inequality and setting $K = (1 + C)\|g\|_{L^1}$, we immediately obtain estimate (2.3.27). □

2.4 Some applications of DREs and BTQFs

In this section, we will apply DREs and BTQFs to rational approximations, eigenvalue problems, and boundary value problems of ordinary differential equations (ODEs). Other applications can be found in the following chapters.

A. Rational approximation

From a special case of Darboux formula (1.1.3) it is easy to obtain (1.1.6), the Obreschkoff formula. If $F = f'$, $b = x$, and $a = x_0$, then the Obreschkoff formula can be rewritten as

$$\begin{aligned}&\sum_{v=0}^{m}\binom{m+k-v}{k}\frac{(x-x_0)^v}{v!}f^{(v)}(x_0)\\ =\ & \sum_{v=0}^{k}(-1)^v\binom{m+k-v}{k}\frac{(x-x_0)^v}{v!}f^{(v)}(x) + R_{m+k},\end{aligned} \tag{2.4.1}$$

where

$$R_{m+k} = -\frac{1}{k!m!}\int_{x_0}^{x}(x-t)^m(x_0-t)^k f^{(m+k+1)}(t)dt.$$

For ease of application, we rewrite equation (2.4.1) in the following form, which was first given by E. Beck.

$$
\begin{aligned}
&\sum_{v=0}^{k}(-1)^v \frac{\binom{k}{v}}{\binom{m+k}{v}} \frac{(x-x_0)^v}{v!} f^{(v)}(x) \\
&= \sum_{v=0}^{m} \frac{\binom{m}{v}}{\binom{m+k}{v}} \frac{(x-x_0)^v}{v!} f^{(v)}(x_0) \\
&+ \frac{1}{(m+k)!} \int_{x_0}^{x} (x-t)^m (x_0-t)^k f^{(m+k+1)}(t)dt. \qquad (2.4.2)
\end{aligned}
$$

We now use formula (2.4.2) without the integral remainder to establish the *rational approximations* for some functions.

Example 2.4.1. In (2.4.2), setting $x_0 = 0$ and $f(x) = e^x$ while dropping the integral remainder generates a rational approximation of the function e^x.

$$
e^x \approx \frac{\sum_{v=0}^{m}(m+k-v)! \binom{m}{v} x^v}{\sum_{v=0}^{k}(-1)^v (m+k-v)! \binom{k}{v} x^v}. \qquad (2.4.3)
$$

In particular, for $m = k$,

$$
e^x \approx \frac{\sum_{v=0}^{k}(2k-v)! \binom{k}{v} x^v}{\sum_{v=0}^{k}(-1)^v (2k-v)! \binom{k}{v} x^v}. \qquad (2.4.4)
$$

We can use formula (2.4.4) to give a sequence of approximation formulas of e^x for $k = 1, 2, \ldots$. For instance, let k be 1, 2, 3, 4, and 5; we have the following rational approximation formulas $a, b, c, d,$ and e, respectively.

$$
a. \ \frac{2+x}{2-x} \qquad b. \frac{12+6x+x^2}{12-6x+x^2}
$$

$$
c. \ \frac{120+60x+12x^2+x^3}{120-60x+12x^2-x^3}
$$

$$
d. \ \frac{1680+840x+180x^2+20x^3+x^4}{1680-840x+180x^2-20x^3+x^4}
$$

$$
e. \ \frac{30240+15120x+3360x^2+420x^3+30x^4+x^5}{30240-15120x+3360x^2-420x^3+30x^4-x^5}
$$

Some numerical results obtained from the above formulas are given in the following table.

x	e^x	a	b	c	d	e
1	2.718281828	3.0	2.714	2.71831	2.71828172	2.718281828
2	7.389056		7.00	7.400	7.38889	7.3890578
3	20.08554		13.00	20.7	20.065	20.08597

By substituting approximation formula (2.4.4) into hyperbolic functions (e.g., $th\ x = (e^{2x} - 1)/(e^{2x} + 1)$), we can derive a class of rational approximation formulas for *hyperbolic functions*.

Example 2.4.2. In formula (2.4.2), we eliminate the integral remainder and set $x_0 = 1$ and $f(x) = x^n, n \in \mathbb{R}$, obtaining

$$\sum_{v=0}^{k}(-1)^v \frac{\binom{k}{v}}{\binom{m+k}{v}} \frac{(x-1)^v}{v!} n(n-1)\cdots(n-v+1)x^{n-v}$$

$$\approx \sum_{v=0}^{m}(-1)^v \frac{\binom{m}{v}}{\binom{m+k}{v}} \frac{(x-1)^v}{v!} n(n-1)\cdots(n-v+1).$$

Therefore,

$$x^n \approx \frac{\displaystyle\sum_{v=0}^{m} \frac{\binom{m}{v}\binom{n}{v}}{\binom{m+k}{v}}(x-1)^v}{\displaystyle\sum_{v=0}^{k}(-1)^v \frac{\binom{k}{v}\binom{n}{v}}{\binom{m+k}{v}} \frac{(x-1)^v}{x^v}}, \tag{2.4.5}$$

or equivalently,

$$x^n \approx \frac{x^k \displaystyle\sum_{v=0}^{m}(m+k-v)!n(n-1)\cdots(n-v+1)\binom{m}{v}(x-1)^v}{\displaystyle\sum_{v=0}^{k}(-1)^v(m+k-v)!n(n-1)\cdots(n-v+1)\binom{k}{v}(x-1)^v x^{k-v}}.$$

In particular, when $m = k$, this formula becomes

$$x^n \approx \frac{x^k \sum_{v=0}^{k} (2k-v)!n(n-1)\cdots(n-v+1)\binom{k}{v}(x-1)^v}{\sum_{v=0}^{k}(-1)^v(2k-v)!n(n-1)\cdots(n-v+1)\binom{k}{v}(x-1)^v x^{k-v}}.$$

In the above approximation, if $k = 2$, $x = 2$, and $n = 1/3$, then

$$\sqrt[3]{2} \approx \frac{\left(12+2-\frac{2}{9}\right)4}{12\cdot 4 - 2\cdot 2 - \frac{2}{9}} = \frac{248}{197} \approx 1.2589.$$

Example 2.4.3. In formula (2.4.2), dropping the integral remainder while setting $x_0 = 1$ and $f(x) = \ln x$ gives

$$\begin{aligned} \ln x \quad &\approx \quad \sum_{v=1}^{m} (-1)^{v-1} \frac{\binom{m}{v}}{\binom{m+k}{v}} \frac{(x-1)^v}{v} \\ &\quad + \sum_{v=1}^{k} \frac{\binom{k}{v}}{\binom{m+k}{v}} \frac{(x-1)^v}{vx^v}. \end{aligned} \tag{2.4.6}$$

For $m = k$, formula (2.4.6) is changed into

$$\ln x \approx \sum_{v=1}^{k} \frac{\binom{k}{v}}{v\binom{2k}{v}} \left[(-1)^{v-1} + \frac{1}{x^v}\right](x-1)^v. \tag{2.4.7}$$

In particular, setting $k = 1$ and $k = 2$, we have, respectively, the following two rational approximation formulas for $\ln x$.

$$\ln x \approx \frac{1}{2}\left(x - \frac{1}{x}\right). \tag{2.4.8}$$

$$\ln x \approx \frac{x^2-1}{12x^2}(8x - x^2 - 1). \tag{2.4.9}$$

Analogous to the Padé approximation, when $m = k$, formula (2.4.2) very often gives a better approximation than when $m \neq k$. In this case, by using a simple variable transform, we can change equation (2.4.2) into

$$\begin{aligned} f(x) &= f(x_0) + \sum_{v=1}^{k} \frac{(2k-v)!}{(2k)!} \binom{k}{v} \\ &\quad \times \left[f^{(v)}(x_0) - (-1)^v f^{(v)}(x) \right] (x - x_0)^v + R_{2k}, \end{aligned} \tag{2.4.10}$$

where

$$R_{2k} = \frac{(x-x_0)^{2k+1}}{(2k)!} \int_0^1 t^k (t-1)^k f^{(2k+1)}(x_0 + t(x - x_0)) dt.$$

Formula (2.4.10) is called the *Hummel–Seebeck–Obrechkoff (HSO) formula* (see [45] and [57]). For the function $f(x) = e^x$, the corresponding rational approximation formula, at $x_0 = 0$, derived from HSO formula (2.4.10) is shown in (2.4.4). It is interesting to note that this approximation is exactly the $[k/k]$ order Padé approximation of e^x. However, we cannot expect the HSO formula to always give a *Padé approximation* for any function. In fact, if $f(x) = e^{\tan^{-1} x}$, then the HSO formula at $x_0 = 0$ for $k = 1$ gives the rational approximation

$$f(x) \approx \frac{1 + \frac{1}{2}x + x^2 + \frac{1}{2}x^3}{1 - \frac{1}{2}x + x^2}.$$

This is different from the [3/2] Padé approximation of the function

$$f(x) \approx \frac{1 + 0.84x + 0.87x^2 + \frac{17}{60}x^3}{1 - 0.16x + 0.53x^2}.$$

A question arises naturally. For what types of functions does the HSO formula give a rational approximation? As Cheney and Southard ([8]) pointed out, as long as a function satisfies the differential equation

$$f'(x) = R_1(x) f(x) + R_2(x) \tag{2.4.11}$$

for arbitrary rational functions $R_1(x)$ and $R_2(x)$, the HSO formula gives a rational approximation of $f(x)$.

There are many functions that satisfy equation (2.4.11). Examples include e^x, $e^{-x^2/2}$, 10^x, $\log_b x$, $\tan^{-1} x$, $(x^2+1)^{-1/2}(a + \sinh^{-1} x)$, $x^{-n}(c + d \ln x)$, $(1-x^2)^{-1/2}$, and $[R(x)]^2$, where a, b, c, and d are any real numbers with $b > 0$ and $b \neq 1$, n is an integer, and $R(x)$ is any rational function.

In the following, we will give a method for constructing a rational approximation for a bivariate function. This procedure is based on DRE (1.3.7) and can also be considered as an extension of the above case of one variable functions.

Let $C_2 = [x_0, x] \otimes [y_0, y]$, $x_0 < x$, $y_0 < y$, be a rectangular region. Denote

$$[f(x, y)]_{x_0}^{x} = f(x, y) - f(x_0, y),$$

and

$$[f(x, y)]_{y_0}^{y} = f(x, y) - f(x, y_0).$$

Assume that in equation (1.3.7) $G(x)$ is a polynomial of degree m in x, with the coefficient of x^m being 1. Substituting $F(x, y) = \partial^2 f/\partial x \partial y$ into (1.3.7) yields

$$f(x, y) = f(x_0, y) + f(x, y_0) - f(x_0, y_0) + \sum_{k=0}^{m-1} \frac{(-1)^k}{m!} \left[\left(\frac{d^{m-k-1} G}{dx^{m-k-1}} \right) \left[\frac{\partial^{k+1} f}{\partial x^{k+1}} \right]_{y_0}^{y} \right]_{x_0}^{x} + R_m, \quad (2.4.12)$$

where the remainder R_m is

$$R_m = \frac{(-1)^m}{m!} \int_{x_0}^{x} G(x) \left[\frac{\partial^{m+1} f}{\partial x^{m+1}} \right]_{y_0}^{y} dx. \quad (2.4.13)$$

In order to have the smallest possible remainder estimate in formula (2.4.12), we choose a Legendre polynomial as $G(x)$:

$$G(x) = \frac{m!}{(2m)!} \left(\frac{d}{dx} \right)^m [x^m (x-1)^m], \quad 0 \le x \le 1.$$

Then, the corresponding DRE (2.4.12) for $0 \le x \le 1$ and $0 \le y \le 1$ is

$$\begin{aligned} f(x, y) &= f(0, y) + f(x, 0) - f(0, 0) \\ &+ \sum_{k=1}^{m} \frac{(-1)^{k-1}}{(2m)!} \left[\left[\frac{\partial^k f}{\partial x^k} \right]_0^y \left(\frac{d}{dx} \right)^{2m-k} (x^2 - x)^m \right]_0^x \\ &+ R_m, \end{aligned} \quad (2.4.14)$$

where R_m has the following two estimates.

$$|R_m| \le \frac{m!}{(2m)!} \left(\frac{2x}{m + 0.5} \right)^{1/2} \left\| \frac{\partial^{m+1} f}{\partial x^{m+1}} \right\|_C. \quad (2.4.15)$$

$$|R_m| \le \frac{m!}{(2m)!} \left(\frac{xy}{2m + 1} \right)^{1/2} \left\| \frac{\partial^{m+2} f}{\partial x^{m+1} \partial y} \right\|_C. \quad (2.4.16)$$

Here, $\|\cdot\|_C$ is the Chebyshev norm of the function over $C_2 = [0,1] \otimes [0,1]$. Since the proofs of (2.4.15) and (2.4.16) are similar, we prove only the former. Applying the Schwartz inequality to the integral in the expression of R_m (see formula (2.4.13) and noting $x_0 = 0$ and $y_0 = 0$, we have

$$
\begin{aligned}
|R_m| &\leq \tfrac{1}{m!}\left(\int_0^x (G(x))^2 dx\right)^{1/2}\left(\int_0^x \left(\left[\tfrac{\partial^{m+1} f}{\partial x^{m+1}}\right]_0^y\right)^2 dx\right)^{1/2} \\
&\leq \tfrac{1}{m!}\left(\tfrac{(m!)^4}{(2m)!(2m+1)!}\right)^{1/2} \cdot 2 \cdot \left\|\tfrac{\partial^{m+1} f}{\partial x^{m+1}}\right\|_C \cdot \sqrt{x} \\
&\leq \frac{2\sqrt{x}m!}{(2m)!\sqrt{2m+1}}\left\|\frac{\partial^{m+1} f}{\partial x^{m+1}}\right\|_C,
\end{aligned}
$$

thereby proving that expression (2.4.15) holds.

It is easy to see that both inequalities (2.4.15) and (2.4.16) give smaller estimates for larger values of m. For instance, for $m = 8$ we obtain $|R_m| \leq \frac{1}{1037836800}\left\|\frac{\partial^9 f}{\partial x^9}\right\|_C$ and $|R_m| \leq \frac{1}{2075673600}\left\|\frac{\partial^{10} f}{\partial x^9 \partial y}\right\|_C$.

We now use formula (2.4.12) to derive rational approximation formulas for some example functions.

Example 2.4.4. Consider the function $\ln(1+x+y)$, $0 \leq x \leq 1$, $0 \leq y \leq 1$. From formula (2.4.12),

$$
\begin{aligned}
\ln(1+x+y) &\approx \ln(1+x) + \ln(1+y) \\
&+ \sum_{k=1}^{m} \frac{(k-1)!}{(2m)!}\left[\left(\frac{1}{(1+x+y)^k} - \frac{1}{(1+x)^k}\right) G_k(x)\right]_0^x . \qquad (2.4.17)
\end{aligned}
$$

As for the one-variable functions $\ln(1+x)$ and $\ln(1+y)$, applying formula (2.4.7) with the replacement of x by $x+1$ gives

$$
\ln(1+x) \approx \sum_{k=1}^{m} \frac{\binom{m}{k}}{k\binom{2m}{k}}\left[\left(\frac{x}{1+x}\right)^k - (-x)^k\right]. \qquad (2.4.18)
$$

Substituting the above expression and the corresponding expression of $\ln(1+y)$ into formula (2.4.17), we thus obtain a bivariate rational approximation for $\ln(1+x+y)$.

We should point out that formula (2.4.18) can be used to construct a rational approximation of $\ln(1+x+y)$ by replacing x with $x+y$ in the expression, but the approximation accuracy of this is generally lower than the accuracy of formula (2.4.17). As an example, set $m = 4$ and $x = y = 1$ (where $\ln(2)$ is found by formula (2.4.18)). Then the approximation value of $\ln(3)$ is

$$
\ln 3 \approx 1.0958445,
$$

with the error 0.00277. If we use HSO type formula (2.4.18) to evaluate the approximation value of $\ln 2$ (by substituting $x = 2$ into formula (2.4.18)), then we obtain

$$\ln 3 \approx 1.0935744$$

with error 0.00504.

Obviously, larger values of m improve the accuracy of formula (2.4.17).

In general, rational approximation formulas constructed from DREs (e.g., DRE (2.4.14)) give better approximations than those derived from HSO expressions. For the DRE formulas, the error from the remainder R_m is especially small. The main error sources are the approximations for $f(x, 0)$ and $f(0, y)$.

Example 2.4.5. We now find the rational approximation of the function $f(x, y) = e^{xy}$. Substituting f into DRE (2.4.14), we have

$$\begin{aligned} e^{xy} \quad &\approx \quad 1 + \sum_{k=1}^{m} \frac{(-1)^{k-1}}{(2m)!} y^k G_k(x) e^{xy} \\ &\quad - \sum_{k=1}^{m} \frac{(-1)^{k-1}}{(2m)!} y^k G_k(0). \end{aligned} \tag{2.4.19}$$

It is not hard to evaluate

$$\begin{aligned} G_k(0) \quad &= \quad \left(\tfrac{d}{dx}\right)^{2m-k} [x^m (x-1)^m] \\ &= \quad (-1)^k \binom{m}{k} (2m-k)!. \end{aligned}$$

Substituting $G_k(0)$ into expression (2.4.19) and solving for e^{xy}, we obtain

$$e^{xy} \approx \frac{(2m)! + \sum_{k=1}^{m} (2m-k)! \binom{m}{k} y^k}{(2m)! + \sum_{k=1}^{m} (-1)^k y^k G_k(x)}. \tag{2.4.20}$$

Clearly, the above formula holds exactly when $x = 0$ or $y = 0$. In addition, if $x = 1$ or $y = 1$, the formula gives a rational approximation for the one-variable function e^y or e^x, respectively.

The formula (2.4.20) is not symmetric about x and y. However, we can derive a symmetric formula by exchanging x and y in formula (2.4.20) and adding the numerator and denominator of the resulting formula to the numerator and denominator of the original formula (2.4.20), respectively.

Similar to the one variable setting, we can construct rational approximation formulas for some bivariate hyperbolic functions by using the rational approximation of bivariate exponential functions.

In formula (2.4.20), taking $x = 1$ and noting that $G_k(1) = (2m-k)!\begin{pmatrix} m \\ k \end{pmatrix}$, we have

$$e^y \approx \frac{(2m)! + \sum_{k=1}^{m}(2m-k)!\begin{pmatrix} m \\ k \end{pmatrix} y^k}{(2m)! + \sum_{k=1}^{m}(2m-k)!\begin{pmatrix} m \\ k \end{pmatrix}(-y)^k}. \tag{2.4.21}$$

It is very interesting that the above approximation expression includes all asymptotic rational expressions of the Euler continued fraction

$$e^y = 1 + \frac{2y}{2-y} \; + \; \frac{y^2}{6} \; + \; \frac{y^2}{10} \; + \; \cdots \; + \; \frac{y^2}{2(2n+1)} \; + \; \cdots .$$

Although formula (2.4.21) can also be found from HSO formula (2.4.19), the HSO formula cannot generate the following, which is a special case of expression (2.4.20) for $y = 1$.

$$e^x \approx \frac{(2m)! + \sum_{k=1}^{m}(2m-k)!\begin{pmatrix} m \\ k \end{pmatrix}}{(2m)! + \sum_{k=1}^{m}(-1)^k G_k(x)}. \tag{2.4.22}$$

Note that the numerator of the above formula does have the variable x. It is worth pursuing further which formula, (2.4.21) or (2.4.22), is better.

Following the above process for different $G(x)$s, we can construct from DRE (1.3.7) many efficient multivariate rational approximation formulas.

B. Eigenvalue problem

In applications, we frequently encounter the following *eigenvalue problem.* Given a linear differential equation with a parameter λ, generally called an eigenvalue, and some homogeneous boundary conditions (besides the trivial solution $y = 0$, these conditions constrain the equation to having solutions for suitable values of λ only), find the smallest value of λ such that the boundary value problem has a solution.

We will use an example to show that the Petr formula, (1.1.5), given in Section 1 is very useful for solving the eigenvalue problem.

Let us consider a second order differential equation

$$y'' + y' + \lambda y = 0 \tag{2.4.23}$$

and the boundary conditions

$$y(0) = 0, \qquad y'(1) = 0.$$

Since a linear homogeneous differential equation has an undetermined amplitude, we can impose any value for $y'(0)$. For example, $y'(0) = 1$.

Under these conditions, equation (2.4.23) uniquely determines the values of the derivatives of y at the point $x = 0$. Let

$$y = a_0 + a_1 x + a_2 x^2 + a_3 x^3 + \cdots .$$

Substitute it into (2.4.23) and then compare the coefficients of the monomials of the same power on the two sides of the equation. We then have

$$\begin{cases} a_0 = 0, a_1 = 1, a_2 = -1/2, a_3 = (1-\lambda)/6, \ldots \\ y(0) = 0, y'(0) = 1, y''(0) = -1, y'''(0) = 1 - \lambda, \ldots . \end{cases} \tag{2.4.24}$$

Assume that

$$x = 1 + \xi, y = b_0 + b_1\xi + b_2\xi^2 + b_3\xi^3 + \cdots .$$

By using the same method, we obtain the values of the derivatives of $y = y(x)$ at another end point, $x = 1$.

$$\begin{cases} b_0, b_1 = 0, \ b_2 = -\lambda b_0/2, \ b_3 = \lambda b_0/6, \ldots \\ y(1) = b_0, \ y'(1) = 0, \ y''(1) = -\lambda b_0, \ y'''(1) = \lambda b_0, \ldots , \end{cases} \tag{2.4.25}$$

where b_0 is unknown.

In the Petr formula, let $F(x)$ be $y''(x)$. From (2.4.24) and (2.4.25), we have $F'(x) = y'''(x)$ and

$$F(0) = -1, \ F(1) = -\lambda b_0 ,$$
$$F'(0) = 1 - \lambda, \ F'(1) = \lambda b_0.$$

To find the expressions of b_0, we set $m = 2$ in formula (1.1.5) and obtain

$$\int_0^1 y''(x)dx = y'(1) - y'(0) = \frac{6(-1 - \lambda b_0) + (1 - \lambda - \lambda b_0)}{12}.$$

Hence,

$$b_0 = (7 - \lambda)/7\lambda. \tag{2.4.26}$$

Similarly, setting $F(x) = y'(x)$ and $m = 3$ in formula (1.1.5) and noting that $F(0) = 1$, $F(1) = 0$, $F'(0) = -1$, $F'(1) = -\lambda b_0$, $F''(0) = 1 - \lambda$, and $F''(1) = \lambda b_0$, we can rewrite equation (1.1.5) as

$$\int_0^1 y'(x)dx = y(1) - y(0) = \frac{60 + 12(-1 + \lambda b_0) + (1 - \lambda + \lambda b_0)}{120}.$$

Therefore,

$$b_0 = (49 - \lambda)/(120 - 13\lambda). \tag{2.4.27}$$

Compare the right sides of equations (2.4.26) and (2.4.27), and we establish a quadratic equation for determining λ:

$$20\lambda^2 - 554\lambda + 840 = 0.$$

The above equation has the two roots

$$\lambda_1 = 1.6098\,, \qquad \lambda_2 = 26.0902. \tag{2.4.28}$$

Of these two roots, the smaller one is more meaningful and stable in the following sense: if more terms of the series solution of differential equation (2.4.23) and larger values of m are involved, then the value of the smaller λ will not change very much (but the value of the larger λ will change substantially). For instance, if we consider the four-term series expansion of the solution y of the differential equation, following the process shown above, we first find $y'''(1)$ and $y'''(0)$ and then make use of formula (1.1.5) for $m = 3$ and $m = 4$, respectively. We thus obtain two other equations of b_0 and λ:

$$b_0 = \frac{71 - 10\lambda}{\lambda(73 - \lambda)}, \qquad b_0 = \frac{679 - 18\lambda}{1680 - 201\lambda + \lambda^2}.$$

Similarly, we end up with a cubic equation for determining the value of λ:

$$28\lambda^3 - 4074\lambda^2 + 80638\lambda - 119280 = 0.$$

Essentially, the above equation can be considered as a quadratic equation because the cubic term, $28\lambda^3$, is not very large. Solving the above equation without the cubic term, denote by λ_0 the smaller solution. Then we replace the cubic term by $28\lambda_0^3$ in the above equation and solve it again. The smallest solution we find is

$$\lambda_1 = 1.608467.$$

(The other two solutions are $\lambda_2 = 21.6694$ and $\lambda_3 = 122.222$.) Comparing this with λ_1 in (2.4.28) shows that they are close. It means that these values of λ_1 are close to the actual value of the minimum of λ. In this sample problem, the actual minimum λ can be found as follows.

$$\lambda = \frac{1}{4} + \theta^2, \tag{2.4.29}$$

where θ is the solution of

$$\tan\theta = 2\theta. \tag{2.4.30}$$

This equation has the minimum solution

$$\theta_{\min} = 1.1655618.$$

Hence,

$$\lambda_{\min} = 1.608534.$$

The approximation error of λ_1 for $m = 2$ and 3 is -0.0010 and the error for $m = 3$ and 4 is 0.000067.

The Ritz method for eigenvalue problems is based on the minimization of an integral functional. As the result, it can be used only for self-adjoint differential operators. The method shown in this section does not need the differential operators to be self-adjoint, and it can even be applied to some nonlinear differential equations.

C. Boundary value problem

From part B of this section, we deduce that the Petr formula can also be used to solve boundary value problems. As an example, we continue to examine the *boundary value problem* shown in part B. To the differential equation (2.4.23), we apply the transform

$$x = ax_1, \tag{2.4.31}$$

where x_1 is a new variable and a is a constant parameter. Equation (2.4.23) is of the form

$$y'' + ay' + a^2\lambda y = 0$$

and, accordingly, tables (2.4.24) and (2.4.25) becomes

$$\begin{cases} a_0 = 0,\ a_1 = a,\ a_2 = -a^2/2,\ a_3 = a^3(1-\lambda)/6, \ldots \\ y(0) = 0,\ y'(0) = a,\ y''(0) = -a^2,\ y'''(0) = a^3(1-\lambda), \ldots \end{cases}$$

and

$$\begin{cases} y(1) = b_0,\ y'(1) = b_1,\ y''(1) = -ab_1 - a^2\lambda b_0, \\ y'''(1) = (1-\lambda)a^2 b_1 + a^3\lambda b_0, \ldots \end{cases}$$

respectively. Here b_0 and b_1 are both unknown. Similar to equation (2.4.26), we have

$$b_1 - a = [6(-a^2 - ab_1 - a^3\lambda b_0) + (1-\lambda)a^3 - (1-\lambda)a^2 b_1 - a^3\lambda b_0]/12,$$

which leads to the linear equation

$$\begin{aligned} & (6a^2\lambda + a^3\lambda)b_0 + [12 + 6a + (1-\lambda)a^2]b_1 \\ = \ & 12a - 6a^2 + (1-\lambda)a^3. \end{aligned} \tag{2.4.32}$$

The second equation about b_0 and b_1 can be found by using formula (1.1.5) for $m = 2$:

$$b_0 = [6(a + b_1) + (-a^2 + ab_1 + a^2\lambda b_0)]/12.$$

The above equation can be rewritten as

$$(12 - a^2\lambda)b_0 - (6 + a)b_1 = 6a - a^2. \tag{2.4.33}$$

Solving equations (2.4.32) and (2.4.33) yields

$$\begin{aligned} b_0 &= \frac{a(144 - 12a^2\lambda)}{144 + 72a + 12a^2(1 + \lambda) + 6a^3\lambda + a^4\lambda^2}, \\ b_1 &= a\frac{144 - 72a + 12a^2(1 - 5\lambda) + 6a^3\lambda + a^4\lambda^2}{144 + 72a + 12a^2(1 + \lambda) + 6a^3\lambda + a^4\lambda^2}. \end{aligned}$$

Returning to the original variable x, since $y(a) = b_0$ and $ay'(a) = b_1$, there are the following two approximation expressions for $y(x)$ and $y'(x)$.

$$\tilde{y}(x) = \frac{x(144 - 12\lambda x^2)}{144 + 72x + 12(1 + \lambda)x^2 + 6\lambda x^3 + \lambda^2 x^4}$$

and

$$\widetilde{y'}(x) = \frac{144 - 72x + 12(1 - 5\lambda)x^2 + 6\lambda x^3 + \lambda^2 x^4}{144 + 72x + 12(1 + \lambda)x^2 + 6\lambda x^3 + \lambda^2 x^4}.$$

The approximation expressions of $y''(x)$ and the higher order derivatives of $y(x)$ can be found by using differential equation (2.4.23) and expressions (2.4) and (2.4).

To check the accuracy of our approximation solution, use $\lambda = -2$ as the example. The corresponding exact solutions are

$$y(x) = (e^x - e^{-2x})/3$$

and

$$y'(x) = (e^x + 2e^{-2x})/3.$$

So the actual values of y and y' at $x = 1$ are

$$y(1) = 0.860982 \quad \text{and} \quad y'(1) = 0.996317.$$

The approximation values calculated using (2.4) and (2.4),

$$\tilde{y}(1) = 0.8571 \quad \text{and} \quad \widetilde{y'}(1) = 1,$$

are indeed very close.

2.5 BTQFs over axially symmetric regions

In this section, we will discuss the algebraic approach to constructing BTQFs for a multiple integral over a bounded closed region Ω in $\mathbb{R}^n$, which is of the form

$$\int_\Omega w(X)f(X)dX.$$

In this expression, $w(X)$ and $f(X)$ are continuous on Ω, and $w(X)$ is the weight function. ($w(X)$ can be 1 particularly.) We are seeking the BTQF of the integral with the form

$$\int_\Omega w(X)f(X)dX \approx \sum_{0\le m_1+\cdots+m_n\le m} \sum_{i\in I} a_i^{m_1,\ldots,m_n} D^{m_1,\ldots,m_n} f(X_i), \qquad (2.5.1)$$

where dX is the volume measure; $a_i^{m_1,\ldots,m_n}$ ($i \in I$ and $0 \le m_1+\cdots+m_n \le m$) are real or complex quadrature coefficients; $D^{m_1,\ldots,m_n} = \partial^{m_1+\cdots+m_n}/\,\partial x_1^{m_1}\cdots x_n^{m_n}$; and $X_i = (x_{i,1}, x_{i,2}, \ldots, x_{i,n})$ ($i \in I$) are evaluation points (or nodes) of f on $\partial\Omega$, the boundary of Ω. In particular, when $m = 0$ we write $a_i^{m_1,\ldots,m_n} = a_i$ and formula (2.5.1) can be rewritten as

$$\int_\Omega w(X)f(X)dX \approx \sum_{i\in I} a_i f(X_i). \qquad (2.5.2)$$

(2.5.2) is called a *BTQF without derivative terms*. When $m \neq 0$, (2.5.1) is called a *BTQF with derivative terms*. The corresponding error functionals of approximations (2.5.1) and (2.5.2) are defined respectively by

$$E(f) \equiv E(f;\Omega) = \int_\Omega w(X)f(X)dX - \sum_{0\le m_1+\cdots+m_n\le m} \sum_{i\in I} a_i^{m_1,\ldots,m_n} D^{m_1,\ldots,m_n} f(X_i) \qquad (2.5.3)$$

and

$$E(f) \equiv E(f;\Omega) = \int_\Omega w(X)f(X)dX - \sum_{i\in I} a_i f(X_i). \qquad (2.5.4)$$

Suppose that $\partial\Omega$ can be described by a system of parametric equations. In particular, the points $X = (x_1, \ldots, x_n)$ on $\partial\Omega$ satisfy the equation (see (1.3.1))

$$\Phi(X) = 0, \qquad (2.5.5)$$

Φ having continuous partial derivatives. In addition, $\Phi(X) \le 0$ for all points in Ω.

Let S be another region in $\mathbb{R}^n$, and let $J : Y = JX$, $X \in \Omega$, be a transform from Ω to S with positive the Jacobian

$$|J| = \left|\frac{\partial(Y)}{\partial(X)}\right| > 0,$$

$X \in \Omega$. J is one-to-one and has the inverse $J^{-1} : X = J^{-1}Y$, $Y \in S$. Denote $w_1(Y) = w_1(JX) = w(X)$. Then for any continuous function $g(X)$

$$\int_S w_1(Y)g(Y)dY = \int_\Omega w_1(Y)g(Y)|J|dX.$$

Denoting $Y_i = JX_i$ $(i \in I)$, $|J_i| = |J|_{X=X_i}$, and taking $f(X) = |J|g(Y) = |J|g(JX)$ in equation (2.5.4), we obtain

$$\begin{aligned} E(|J|g; \Omega) &= \int_\Omega w(X)|J|g(Y)dX - \sum_{i\in I} a_i|J_i|g(Y_i) \\ &= \int_S w_1(Y)g(Y)dY - \sum_{i\in I} b_i g(Y_i), \end{aligned}$$

where $b_i = a_i|J_i|$ $(i \in I)$. Obviously, if Y, the boundary points of S, satisfy $\Phi_1(Y) = \Phi_1(JX) = \Phi(X) = 0$, then J maps the boundary evaluation points X_i $(i \in I)$ on Ω onto the boundary evaluation points $Y_i = JX_i$ on S. Consequently, we have the following result.

Theorem 2.5.1 *Let the error functional of the quadrature formula*

$$\int_S w_1(Y)g(Y)dY \approx \sum_{i\in I} b_i g(Y_i) \tag{2.5.6}$$

be

$$E(g; S) = \int_S w_1(Y)g(Y)dY - \sum_{i\in I} b_i g(Y_i).$$

Then

$$E(g; S) = E(|J|g; \Omega).$$

In particular, if $|J|$ *is a constant, then* $E(g; S) = |J|E(g; \Omega)$. *In this case,* $E(g; \Omega) = 0$ *implies* $E(g; S) = 0$.

In addition, if the boundary of S is defined by $\Phi_1(Y) = \Phi_1(JX) = \Phi(X) = 0$ *and* $\Phi(X) = 0$ *defines the boundary of* Ω, *then quadrature formula (2.5.6) is also a BTQF.*

In this section, we will establish the BTQFs over some special regions. Theorem 2.5.1 tells us that we can construct the BTQFs over many more regions from the BTQFs over the special regions by using certain transforms. In addition, if the transform is linear, then the new BTQF is of the same algebraic precision degree as the old BTQF because the transform Jacobian is a constant.

Three questions arise during the construction of BTQFs (2.5.1):

(i) What is the highest possible degree of algebraic precision?

(ii) What is the fewest evaluation points needed to construct a BTQF with the highest possible degree of algebraic precision?

(iii) How to construct the *BTQF with the fewest evaluation points* and the highest possible degree of algebraic precision?

We now answer the first question. In most cases, BTQF (2.5.1) has an inherent highest degree of algebraic precision. For instance, if $\Phi(X)$ is a polynomial of degree m, then the highest possible degree of algebraic precision of the BTQF without derivative terms (i.e., formula (2.5.2)) cannot exceed $m-1$ because the summation on the right-hand side of (2.5.2) becomes zero and the integral value on the left-hand side is negative when $f = \Phi$. Hence, when the boundary function Φ is a polynomial of a low degree, to raise the degrees of algebraic precision of the quadrature formulas, we must construct BTQFs with derivative terms (i.e., formula (2.5.1)) with $m \neq 0$.

In the following, we are going to find the solutions to questions (ii) and (iii). To simplify our discussion, we limit the region in question, Ω, to be axially symmetric or fully symmetric. An *axially symmetric region* is a region that, for any point $X = (x_1, \ldots, x_n)$ in it, must contain all points with the form $(\pm x_1, \ldots, \pm x_n)$. The set of axially symmetric points associated with X forms a reflection group. If a region containing a point $X = (x_1, \ldots, x_n)$ also contains all points $(\pm a_1, \ldots, \pm a_n)$, where $(a_1, \ldots, a_n)$ is a permutation of $(x_1, \ldots, x_n)$, then the region is called a *fully symmetric region*. Throughout, we will denote all fully symmetric points, $(\pm a_1, \ldots, \pm a_n)$, associated with X by X_{FS} and call X the generator of the fully symmetric point set. The cardinal number of the set of fully symmetric points associated with a generator $X \in \mathbb{R}^n$ is $2^n(n!)$. Obviously, a fully symmetric region is an axially symmetric region, but the converse is not true.

A quadrature formula is called a *fully symmetric quadrature formula* if the quadrature sum can be divided into several subsums such that in each of the subsums, the evaluation points are fully symmetric and the corresponding quadrature coefficients are the same. In addition, if the fully symmetric evaluation points are on the boundary of the integral region, then the corresponding quadrature formula is called a *fully symmetric BTQF*.

Denote a monomial in terms of X by X^α $(\alpha \in \mathbb{Z}_0^n)$, which can be written in the form

$$X^\alpha = x_1^{\alpha_1}, \ldots, x_n^{\alpha_n},$$

where $(\alpha_1, \ldots, \alpha_n)$ is called the exponent of X^α.

From the definition of the fully symmetric region, we immediately have the following results.

Theorem 2.5.2 *The value of a multiple integral of a monomial X^α over an axially symmetric region is zero if α contains an odd component. The value of a multiple integral of X^α over a fully symmetric region depends on α, but is independent of the order of α_i $(i = 1, \dots, n)$.*

Theorem 2.5.3 *Denote by $\pi_r^n(X)$ the set of all polynomials of degree no less than r. Let Ω be a fully symmetric region,*

$$\int_\Omega f(X)dX \approx \sum_{i\in I} a_i f(X_i) \tag{2.5.7}$$

be a fully symmetric BTQF, and $E : f \to \mathbb{R}$ be the error operator defined by

$$E(f) \equiv E(f;\Omega) = \int_\Omega f(X)dX - \sum_{i\in I} a_i f(X_i).$$

(The above expression is a special form of (2.5.4) with $w(X) = 1$.) Then $\pi_{2k+1}^n \subset N(E)$, the null space of E, if and only if

$$x_1^{2k_1}\cdots x_n^{2k_n} \in N(E) \quad 0 \le k_1 \le \cdots \le k_n,\ k_1 + \cdots + k_n \le k. \tag{2.5.8}$$

Theorem 2.5.3 can be considered as the general principle for constructing fully symmetric BTQFs. First, we set one or more sets of fully symmetric evaluation points, with possibly some unknown points $\{X_i\}$, on the boundary $\partial\Omega$ and assume the quadrature coefficients a_i corresponding to each set to be the same. Then substituting all $f(X) = x_1^{2k_1}\cdots x_n^{2k_n}$ $(0 \le k_1 \le \cdots \le k_n$ and $k_1 + \cdots + k_n \le k)$ into $E(f) = 0$, we obtain a system about X_i and a_i. Finally, we solve the system for X_i and a_i and construct a quadrature formula (2.5.7). However, a fully symmetric quadrature formula usually has too many evaluation points. (Remember that for a point $X \in \mathbb{R}^n$ there are, in general, $2^n(n!)$ fully symmetric points.) In order to reduce the number of evaluation points in the quadrature formula, we can use an alternative form of Theorem 2.5.3 to construct a different type of symmetric quadrature formulas. We will use the following example to illustrate the idea.

Example 2.5.1. Consider a triple integral over the region $C_3 = [-1, 1]^3$. Obviously, the inherent highest degree of algebraic precision of the BTQF is 5. To construct a fully symmetric BTQF, we make use of the following fully symmetric evaluation points.

$$(1,0,0)_{FS}, \quad (1,1,0)_{FS}, \quad \text{and} \quad (1,x_0,x_0)_{FS},$$

where x_0 $(0 < x_0 < 1)$ is undetermined. The three sets of fully symmetric points contain a total of 42 points (6, 12, and 24 points for the first, second, and third set respectively). Let the respective quadrature coefficients for each set of fully symmetric points be L, M, and N, all of which can be found using the general principle for constructing fully symmetric BTQFs. Substitute $f = 1, x^2, x^4$, and x^2y^2 into

$$\int_{C_3} f(x,y,z)dxdydz = a_1 \sum f_6 + a_2 \sum f_{12} + a_3 \sum f_{24},$$

where $\sum f_6$, $\sum f_{12}$, and $\sum f_{24}$ are the sums of the function values of f over the first, second, and third set of symmetric points, respectively. The following system is produced.

$$\begin{cases} 6a_1 + 12a_2 + 24a_3 = 8 \\ 2a_1 + 8a_2 + (8 + 16x_0^2)a_3 = 8/3 \\ 2a_1 + 8a_2 + (8 + 16x_0^4)a_3 = 8/5 \\ 4a_2 + (16x_0^2 + 8x_0^4)a_3 = 8/9. \end{cases}$$

Solving the above system yields

$$x_0 = \sqrt{\frac{5}{8}}, \quad a_1 = \frac{364}{225}, \quad a_2 = -\frac{160}{225}, \quad a_3 = \frac{64}{225},$$

giving the following BTQF of algebraic precision order 5.

$$\int_{C_3} f(x,y,z)dxdydz \approx \frac{4}{225}\left[91 \sum f_6 - 40 \sum f_{12} + 16 \sum f_{24}\right]. \tag{2.5.9}$$

Quadrature formula (2.5.9), first given by Sadowsky [60], uses too many evaluation points. Carefully considering Theorem 2.5.3, we find that the principle of constructing fully symmetric BTQFs shown in the theorem can be used to construct some "partial" symmetric BTQFs with fewer evaluation points.

We first define a new term. A set of points $X_i \in \mathbb{R}^n$ $(i \in I)$ is called a *symmetric point set of degree* k if it possesses the following two properties.

(a) $\sum_{i \in I} f(X_i) = 0$ for all $f(X) = X^\alpha$, where α contains an odd component.

(b) $\sum_{i \in I} f(X_i)$ are the same for all $f(X) = x_1^{2k_1} \cdots x_n^{2k_n}$, $2(k_1 + \cdots + k_n) = r$. Here, $r \le k$.

Obviously, a set of fully symmetric points must be a set of symmetric points of any degree, but the converse is not true. For instance, a symmetric point set of degree 5 may not be a fully symmetric point set. We now list all symmetric point sets of degree 5 on the boundary of C_3 as follows.

$I = \{(\pm 1, \pm x_0, 0), (\pm x_0, 0, \pm 1), (0, \pm 1, \pm x_0), 0 < x_0 < 1\}$

$II = \{(\pm y_0, \pm 1, 0), (\pm 1, 0, \pm y_0), (0, \pm y_0, \pm 1), 0 < y_0 < 1\}$

$$III = \{(\pm 1, \pm 1, 0), (\pm 1, 0, \pm 1), (0, \pm 1, \pm 1)\}$$
$$IV = \{(\pm 1, \pm 1, \pm 1)\}$$
$$V = \{(1, 0, 0)_{FS}\}$$
$$VI = \{(1, x_1, x_2)_{FS}, 0 < x_1, x_2 < 1\}$$
$$VII = \{(1, 1, x_3)_{FS}, 0 < x_3 < 1\}$$

Sets V, VI, and VII are fully symmetric, but others are not.

If a BTQF constructed by using symmetric point set of degree k satisfies condition (2.5.8), then it is called a *symmetric BTQF of degree k*.

Example 2.5.2. As an example, we now use the sets I, III, and IV to construct a symmetric BTQF of degree 5 with 32 evaluation points over C_3. Denote the quadrature coefficients corresponding to I, III, and IV as a_1, a_2, and a_3 respectively.

Following the procedure shown in Example 2.5.1, we obtain the system

$$\begin{cases} 12a_1 + 12a_2 + 8a_3 = 8 \\ 4(1 + x_0^2)a_1 + 8a_2 + 8a_3 = 8/3 \\ 4(1 + x_0^4)a_1 + 8a_2 + 8a_3 = 8/5 \\ 4x_0^2 a_1 + 4a_2 + 8a_3 = 8/9. \end{cases}$$

It has the solution

$$x_0 = \sqrt{\frac{3}{10}}, \quad a_1 = \frac{80}{63}, \quad a_2 = -\frac{52}{63}, \quad a_3 = \frac{1}{3}.$$

Thus, we obtain a symmetric BTQF of degree 5

$$\int_{C_3} f(x, y, z)dxdydz$$
$$\approx \frac{1}{63}\left[80 \sum f_{12}(I) - 52 \sum f_{12}(III) + 21 \sum f_8(IV)\right], \quad (2.5.10)$$

$\sum f_{12}(I)$, $\sum f_{12}(III)$, and $\sum f_8(IV)$ are the sums of the function values of f over the symmetric point sets I, III, and IV, respectively; the numbers in the sub-indices are the cardinal numbers of each set.

Similarly, we can use sets II, III, and IV to construct another symmetric BTQF of degree 5.

$$\int_{C_3} f(x, y, z)dxdydz$$
$$\approx \frac{1}{63}\left[80 \sum f_{12}(II) - 52 \sum f_{12}(III) + 21 \sum f_8(IV)\right], \quad (2.5.11)$$

where $y_0 = \sqrt{\frac{3}{10}}$ in set II.

Quadratures (2.5.10) and (2.5.11) can be considered as two special cases of the following symmetric BTQF of degree 5, which is constructed by using I, II, and IV.

$$\int_{C_3} f(x,y,z)dxdydz \approx \frac{4(1+y_0^2)}{9(y_0^2-x_0^2)} \sum f_{12}(I)$$
$$+\frac{4(1+x_0^2)}{9(x_0^2-y_0^2)} \sum f_{12}(II) + \frac{1}{3} \sum f_8(IV), \qquad (2.5.12)$$

where

$$\sqrt{\frac{3}{10}} \le y \le 1, \; y_0 \ne \sqrt{\sqrt{\frac{13}{5}} - 1}, \text{ and } x_0 = \sqrt{\frac{8-5y_0^2}{5(1+y_0^2)}}.$$

When $y_0 = 1$ and $y_0 = \sqrt{\frac{3}{10}}$ we obtain formulas (2.5.10) and (2.5.11), respectively.

It can be proved that the minimum number of evaluation points of symmetric BTQFs is 32. Since the quadrature formula is symmetric, on each boundary plane we must have the same number of evaluation points. Let the number of evaluation points on each boundary plane be $k = 2$ (obviously, k cannot be 1). The symmetric point set has to be I or II. It is easy to check that the sets cannot yield a symmetric BTQF of degree 5. Similarly, for the cases of $k = 3, \dots, 9$, no matter which symmetric point sets are chosen from $\{I, \dots, VII\}$, we find that there does not exist any symmetric BTQFs of degree 5 with evaluation points less than 32. For $k \ge 10$, every symmetric BTQF of degree 5, if it exists, must have more than 32 evaluation points. Thus, we obtain the following proposition.

Proposition 2.5.4 *There exist infinitely many symmetric BTQFs of degree* 5 *with* 32 *evaluation points. In addition, the number of evaluation points of a symmetric BTQF of degree* 5 *can be no less than* 32.

For BTQF of degree 3, the minimum is reduced to 6 evaluation points. As an example, we give the formula

$$\int_{C_3} f(x,y,z)dxdydz \approx \frac{4}{3}[f(1,0,0) + f(-1,0,0)$$
$$+f(0,1,0) + f(0,-1,0) + f(0,0,1) + f(0,0,-1)].$$

Example 2.5.3. We will use a double-layered spherical shell as an example to demonstrate the techniques of regrouping evaluation points to obtain the symmetric BTQF with the fewest evaluation points. A double-layered spherical shell in $\mathbb{R}^n$, denoted by Sh_n, is defined by

$$Sh_n = \{X \in \mathbb{R}^n : a^2 \le |X| \le b^2\}.$$

It is easy to find that the largest degree of algebraic precision of BTQFs over Sh_n without derivatives is 3. We choose the following point sets as evaluation points.

$VIII = \{(\pm b, 0, \dots, 0), (0, \pm b, 0, \dots, 0), \cdots, (0, \dots, 0, \pm b, 0)\}$

$IX = \{(0, \dots, 0, \pm b)\}$

$X = \{(0, \dots, 0, \pm a)\}$

Obviously, these sets are neither fully symmetric point sets nor symmetric point sets of degree 3, but by using these sets, we can construct a BTQF of degree 3 over Sh_n with the fewest evaluation points. Denote the quadrature coefficients corresponding to VIII, IX, and X by a_1, a_2, and a_3, respectively. The BTQF generated,

$$\int_{Sh_n} f(X)dX \approx a_1 \sum f_{2(n-1)}(VIII) + a_2 \sum f_2(IX) + a_3 \sum f_2(X), \tag{2.5.13}$$

is of algebraic precision of degree 3 if it holds exactly for $f = 1, x_1^2$, and x_n^2; i.e., coefficients a_i $(i = 1, 2, 3)$ have to satisfy the system

$$\begin{cases} 2(n-1)a_1 + 2a_2 + 2a_3 &= \pi^{n/2}(b^n - a^n)/\Gamma\left(\frac{n}{2}+1\right) \\ 2b^2 a_1 &= \pi^{n/2}(b^{n+2} - a^{n+2})/\left((n+2)\Gamma\left(\frac{n}{2}+1\right)\right) \\ \qquad 2b^2 a_2 + 2a^2 a_3 &= \pi^{n/2}(b^{n+2} - a^{n+2})/\left((n+2)\Gamma\left(\frac{n}{2}+1\right)\right). \end{cases}$$

Solving the system yields

$$a_1 = \alpha(b^2 - a^2)\left(b^{n+2} - a^{n+2}\right),$$

$$a_2 = \alpha\left(b^{n+4} + (n+1)a^{n+2}b^2 - 3b^{n+2}a^2 - (n-1)a^{n+4}\right),$$

$$a_3 = \alpha b^2\left(2b^{n+2} - (n+2)a^n b^2 + n a^{n+2}\right),$$

where

$$\alpha = \frac{\pi^{n/2}}{2b^2\Gamma\left(\frac{n}{2}+1\right)(n+2)(b^2 - a^2)}.$$

When $n = 2$ and 3, formula (2.5.13) gives BTQFs over a ring domain and a 3-dimensional double-layered spherical shell respectively as follows.

$$\begin{aligned} &\int_{Sh_2} f(x, y)dxdy \\ \approx\ &\frac{\pi(b^2 - a^2)}{8b^2}\Big\{(b^2 + a^2)[f(b, 0) + f(-b, 0)] + 2b^2[f(0, a) + f(0, -a)] \\ &+ (b^2 - a^2)[f(0, b) + f(0, -b)]\Big\}, \end{aligned}$$

$$\int_{Sh_3} f(x,y,z)dxdydz$$
$$\approx \frac{2\pi}{15b^2(b^2-a^2)}\Big\{(b^2-a^2)(b^5-a^5)[f(b,0,0)+f(-b,0,0)+f(0,b,0)$$
$$+f(0,-b,0)]+b^2(2b^5-5a^3b^2+3a^5)[f(0,0,a)+f(0,0,-a)]$$
$$+(b^7-3a^2b^5+4a^5b^2-2a^7)[f(0,0,b)+f(0,0,-b)]\Big\}.$$

Taking the limit $a \to 0$, from quadrature formula (2.5.13) we obtain the following quadrature formula over the sphere S_3, which has the algebraic precision of degree 3.

$$\int_{S_3} f(x,y,z)dxdydz \approx \frac{\pi^{n/2}b^n}{2(n+2)\Gamma\left(\frac{n}{2}+1\right)}$$
$$\times\left(\sum f_{2(n-1)}(VIII)+\sum f_2(IX)+4f(0,\ldots,0)\right).$$

We now prove that BTQF (2.5.13) is a formula with the fewest evaluation points.

Theorem 2.5.5 *The minimum number of evaluation points of BTQFs over an n-dimensional double-layered spherical shell Sh_n is $2(n+1)$. In particular, the minimum number of evaluation points for BTQFs over a ring domain and a 3-dimensional double-layered spherical shell are respectively 6 and 8.*

Proof. For a BTQF over Sh_n with precision degree 3, we will first prove that the minimum of evaluation points on the outside layer of Sh_n cannot be less than $2n$. Without a loss of generality, we assume that the number of evaluation points on the outside layer is $2n-1$. (The cases when the minimums are less than $2n-1$ can be proved similarly.) We will see that a contradiction results from this assumption. If the assumption is valid, we take the limit $a \to 0$ to the BTQF and obtain a quadrature formula over an n-dimensional sphere with $2n-1$ evaluation points as follows.

$$\int_{S_n} f(X)dX \approx a_0 f(0,\ldots,0)+\sum_{i=1}^{2n-1} a_i f(X_i), \tag{2.5.14}$$

where X_i $(i=1,\ldots,2n-1)$ lie on the sphere surface and $a_i \neq 0$ $(i=1,\ldots,n)$. We will prove it cannot be of algebraic precision degree 3.

Let us consider the following $2n$ complex vectors

$$AX_1, \ldots, AX_n, AX_1^2, AX_2^2, \ldots, AX_n^2, \tag{2.5.15}$$

where

$$AX_i = (\sqrt{a_1}x_{1,i}, \sqrt{a_2}x_{2,i}, \ldots, \sqrt{a_{2n-1}}x_{2n-1,i})$$

and

$$AX_i^2 = (\sqrt{a_1}x_{1,i}^2, \sqrt{a_2}x_{2,i}^2, \ldots, \sqrt{a_{2n-1}}x_{2n-1,i}^2).$$

Assume that there exist constants b_i $(i = 1, \ldots, 2n)$ such that

$$\begin{aligned} &b_1 AX_1 + \cdots + b_n AX_n \\ &+b_{n+1}AX_1^2 + b_{n+2}AX_2^2 + \cdots + b_{2n}AX_n^2 = 0. \end{aligned} \tag{2.5.16}$$

Taking dot product with AX_i $(i = 1, \ldots, n)$ on both sides of (2.5.16) and noting that the quadrature sums in (2.5.14) are vanishing for all $f = X^\alpha$ if α has an odd component and $|\alpha| \le 3$, we obtain

$$b_i AX_i \cdot AX_i = b_i \sum_{i=1}^{2n-1} a_i x_i^2 = 0, \ \ i = 1, \ldots, n.$$

Since the sums in the above equation are the quadrature sums of BTQF (2.5.14) for $f(X) = X^\alpha$ with $\alpha = 2\mathbf{e}_i$ ($\{\mathbf{e}_1, \mathbf{e}_2, \ldots, \mathbf{e}_n\}$ being the standard basis of $\mathbb{R}^n$), which are not zero, we obtain $b_i = 0$ for all $i = 1, \ldots, n$. Consequently, equation (2.5.16) is reduced to

$$b_{n+1}AX_1^2 + b_{n+2}AX_2^2 + \cdots + b_{2n}AX_n^2 = 0. \tag{2.5.17}$$

Taking the dot product with $A = (\sqrt{a_1}, \ldots, \sqrt{a_{2n-1}})$ on both sides of equation (2.5.17) and noting that the quadrature sums in (2.5.14) are vanishing for all $f = X^\alpha$ if $\alpha = 3$, we obtain

$$\|\sum_{i=1}^{n} \sqrt{b_{n+i}} AX_i\|_{\ell_2}^2 = 0.$$

Hence,

$$\sqrt{b_{n+1}}AX_1 + \cdots + \sqrt{b_{2n}}AX_n = 0.$$

Similarly, we have $b_{n+i} = 0$ for all $n = 1, \ldots, n$. Thus, vectors (2.5.15) are linearly independent, but this is impossible because all of them have $2n - 1$ components. This contradiction means that the number of evaluation points on the outside layer for any BTQFs over Sh_n with algebraic precision degree 3 must be more than $2n-1$.

We now prove that the number of evaluation points on the inside layer for any BTQFs over Sh_n with precision degree 3 cannot be less than 2. Otherwise, if there is none or there is only one evaluation point, $X_0 = (x_{0,1}, x_{0,2}, \ldots, x_{0,n})$, on the inside layer of Sh_n, then a BTQF over Sh_n with algebraic precision degree 3 is not exact for quadratic polynomial

$$f(X) = \sum_{i=1}^{n} x_i^2 - b^2$$

or for a cubic polynomial

$$f(X) = \left(\sum_{i=1}^{n} x_i^2 - b^2\right)(x_j - x_{0,j}),$$

where $x_{0,j} \neq 0$. This completes the proof of the theorem. □

A similar argument of the proof of Theorem 2.5.5 can be applied to solve other minimum evaluation point problems. For instance, we have the following result.

Theorem 2.5.6 *The minimum number of evaluation points needed for constructing a quadrature formula over an axially symmetric region in $\mathbb{R}^n$ with algebraic precision degree* 3 *is* $2n$.

The construction of a quadrature formula of this type can be found in Section 3.9 of Stroud [67].

To improve the algebraic precision degrees of BTQR's, we use the derivatives of the integrands. As examples, we will construct symmetric quadrature formulas over the surfaces of the regions $C_2 = [-1, 1]^2$, $C_3 = [-1, 1]^3$, and the n-dimensional sphere S_n.

Example 2.5.4. Denote the sets of fully symmetric points $XI = \{(1, 1)_{FS}\}$ and $XII = \{(1, 0)_{FS}\}$. We construct a symmetric BTQF with precision degree 5 over $C_2 = [-1, 1]$ as follows.

$$\begin{aligned}\int_{C_2} f(x, y)dxdy &\approx a_1 \sum f_4(XI) + a_2 \sum f_4(XII)\\ &+a_3[f'_x(1, 1) - f'_x(-1, -1) + f'_x(1, -1) - f'_x(-1, 1)\\ &+f'_y(1, 1) - f'_y(-1, -1) + f'_y(-1, 1) - f'_y(1, -1)]\\ &+a_4[f'_x(1, 0) - f'_x(-1, 0) + f'_y(0, 1) - f'_y(0, -1)].\end{aligned}$$

Obviously, the above quadrature formula is of precision degree 5 if it is exact for

$f(x, y) = 1, x^2, x^4$, and x^2y^2. Therefore, we obtain the following system.

$$\begin{cases} 4a_1 + 4a_2 = 4 \\ 4a_1 + 2a_2 + 8a_3 + 4a_4 = 4/3 \\ 4a_1 + 2a_2 + 16a_3 + 8a_4 = 4/5 \\ 4a_1 + 16a_3 = 4/9. \end{cases}$$

The solution is

$$a_1 = -\frac{1}{15}, \quad a_2 = \frac{16}{15}, \quad a_3 = \frac{2}{45}, \quad a_4 = -\frac{2}{9}.$$

Similarly, we can construct a BTQF over $C_3 = [-1, 1]^3$ with precision degree 7 and 50 fully symmetric evaluation points $XIII = \{(1, 1, 1)_{FS}\}$, $XIV = \{(1, 0, 0)_{FS}\}$, $XV = \{(1, \frac{1}{2}, 0)_{FS}\}$, and $XVI = \{(1, 1, 0)_{FS}\}$ as follows.

$$\int_{C_3} f(x, y, z)dxdydz \approx a_1 \sum f_8(XIII) + a_2 \sum f_6(XIV)$$
$$+a_3 \sum f_{24}(XV) + a_4 \sum f_{12}(XVI) + a_5 M_1 + a_6 M_2 + a_7 M_3,$$

where $a_1 = \frac{1}{5}$, $a_2 = -\frac{16}{105}$, $a_3 = \frac{512}{945}$, $a_4 = -\frac{64}{135}$, $a_5 = -\frac{11}{405}$, $a_6 = -\frac{16}{81}$, $a_7 = \frac{172}{2835}$,

$$\begin{aligned} M_1 \;=\; & f_x'(1, 1, 1) - f_x'(-1, -1, -1) + f_x'(1, 1, -1) - f_x'(-1, -1, 1) \\ & +f_x'(1, -1, -1) - f_x'(-1, 1, 1) + f_x'(1, -1, 1) - f_x'(-1, 1, -1) \\ & +f_y'(-1, 1, -1) - f_y'(1, -1, 1) + f_y'(-1, 1, 1) - f_y'(1, -1, -1) \\ & +f_y'(1, 1, -1) - f_y'(-1, -1, 1) + f_y'(1, 1, 1) - f_y'(-1, -1, -1) \\ & +f_z'(-1, 1, 1) - f_z'(1, -1, -1) + f_z'(1, -1, 1) - f_z'(-1, 1, -1) \\ & +f_z'(1, 1, 1) - f_z'(-1, -1, -1) + f_z'(-1, -1, 1) - f_z'(1, 1, -1), \end{aligned}$$

$$\begin{aligned} M_2 \;=\; & f_x'(1, 0, 0) - f_x'(-1, 0, 0) + f_y'(0, 1, 0) - f_y'(0, -1, 0) \\ & +f_z'(0, 0, 1) - f_z'(0, 0, -1), \end{aligned}$$

and

$$\begin{aligned} M_3 \;=\; & f_x'(1, 1, 0) - f_x'(-1, -1, 0) + f_x'(1, -1, 0) - f_x'(-1, 1, 0) \\ & +f_x'(1, 0, 1) - f_x'(-1, 0, -1) + f_x'(1, 0, -1) - f_x'(-1, 0, 1) \\ & +f_y'(0, 1, 1) - f_y'(0, -1, -1) + f_y'(0, 1, -1) - f_y'(0, -1, 1) \\ & +f_y'(1, 1, 0) - f_y'(-1, -1, 0) + f_y'(-1, 1, 0) - f_y'(1, -1, 0) \\ & +f_z'(1, 0, 1) - f_z'(-1, 0, -1) + f_z'(-1, 0, 1) - f_z'(1, 0, -1) \\ & +f_z'(0, 1, 1) - f_z'(0, -1, -1) + f_z'(0, -1, 1) - f_z'(0, 1, -1). \end{aligned}$$

Example 2.5.5. Choose $2n$ fully symmetric evaluation points $XVII = \{(r, 0, \ldots, 0)_{FS}\}$. We can obtain a BTQF over $S_n(\sum_{i=1}^n x_i^2 \le r^2)$ with the precision degree 3 as follows.

$$\int_{S_n} f(X)dX \approx \frac{\pi^{n/2} r^{n+1}}{2n(n+2)\Gamma\left(\frac{n}{2}+1\right)}\left[\frac{n+2}{r}\sum f_{2n}(XVII) - f'_{x_1}(r, 0, \ldots, 0) + f'_{x_1}(-r, 0, \ldots, 0) - \cdots - f'_{x_n}(0, \ldots, 0, r) + f'_{x_n}(0, \ldots, 0, -r)\right].$$

At the end of this section, we introduce some recent results on the numerical quadrature formulas over $\bar{S}_n = \{X \in \mathbb{R}^n |X| = 1\}$, the surface of the unit sphere $B_n = B_n(1) = \{X \in \mathbb{R}^n |X| \le 1\}$ in $\mathbb{R}^n$. Let H be a function defined on $\mathbb{R}^n$ that is symmetric with respect to x_n; i.e., $H(X, x_n) = H(X, -x_n)$, $X \in \mathbb{R}^{n-1}$. Then for any continuous function f defined on $\bar{S}_n$,

$$\begin{aligned}
&\int_{\bar{S}_n} f(Y)H(Y)d\mu_n \\
&= \int_{B_{n-1}} \left[f\left(X, \sqrt{1-|X|^2}\right) + f\left(X, -\sqrt{1-|X|^2}\right)\right] \\
&\quad \times H(X, \sqrt{1-|X|^2})\frac{dX}{\sqrt{1-|X|^2}},
\end{aligned} \tag{2.5.18}$$

where $Y \in \bar{S}_n$, $X \in \mathbb{R}^{n-1}$, $-1 \le t \le 1$, and $d\omega_n$ is the surface measure on $\bar{S}_n$. The volume of $\bar{S}_n$ is

$$\omega_n = \int_{\bar{S}_n} d\mu_n = \frac{2\pi^{n/2}}{\Gamma\left(\frac{n}{2}\right)}.$$

Formula (2.5.18), shown in Xu [77], can be proved straightforwardly by substituting $d\mu_n = (1-t^2)^{(n-3)/2} dt d\mu_{n-1}$ and $Y = (\sqrt{1-t^2}X, t)$ into the left-hand integral of the equation.

(2.5.18) changes a boundary integral into an integral over the interior of the boundary. Hence it can be used to derive a BTQF over B_n from a quadrature formula of an integral over B_{n-1}. Following [77], suppose that there is a quadrature formula of precision degree m on B_{n-1},

$$\int_{B_{n-1}} g(X)H\left(X, \sqrt{1-|X|^2}\right)\frac{dX}{\sqrt{1-|X|^2}} \approx \sum_{i=1}^{N} a_i g(X_i);$$

that is, the quadrature formula is exact for all polynomials in π_m^{n-1}, which denotes the set of all polynomials defined in $\mathbb{R}^{n-1}$ with a total degree not more than m.

Then there is a quadrature formula of homogeneous precision degree m on $\bar{S}_n$:

$$\int_{\bar{S}_n} f(Y)H(Y)d\mu_n$$

$$\approx \sum_{i=1}^{N} a_i \left[f\left(X_i, \sqrt{1-|X_i|^2}\right) + f\left(X_i, -\sqrt{1-|X_i|^2}\right)\right]. \quad (2.5.19)$$

Recently, Mhaskar, Narcowich, and Ward (see[55]) developed a new method for obtaining quadrature formulas on $\bar{S}_n$, which can be applied to the right-hand integrals of equation (1.3.19) so that the BTQFs over B_n can be constructed.

Let $C \subset \bar{S}_n$ be a set of scattered points, and let R be a finite collection of closed, nonoverlapping (i.e., having no common interior points) regions $E \subset \bar{S}_n$ such that $\cup_{E \in R} = \bar{S}_n$. R is said to be a C-compatible partition if each $E \in R$ is a spherical simplex containing at least one point of C in its interior. The partition norm is defined by $\|R\| = \max_{E \in R} diam\ R$. If the C-compatible partition R has a small enough norm, [55] found a process for constructing quadrature formulas on $\bar{S}_n$, with respect to R, that are exact for spherical harmonics of a fixed order, have nonnegative quadrature coefficients, and are based on the integrand function values at the points in C.

Chapter 3

The Integration and DREs of Rapidly Oscillating Functions

In this chapter, we will discuss the following n-dimensional *oscillatory integral*

$$\int_0^1 f(x_1, \ldots, x_n; \langle N_1 x_1 \rangle, \ldots, \langle N_n x_n \rangle) dx_1 \cdots dx_n,$$

where, for all $i = 1, \ldots, n$, $\langle N_i x_i \rangle = N_i x_i - [N_i x_i]$, the fractional part of $N_i x_i$, and $N_i \geq 2$ are large positive integers. Here, function f in the integral is called a *rapidly oscillating function.* In this chapter, we will show that an integral of a continuous function $f(x_1, \ldots, x_n; y_1, \ldots, y_n)$ over a $2n$-dimensional sphere or cube can be approximated by a sequence of oscillatory integrals. In particular, if the continuous function f is also periodic in terms of each independent variable in a subset consisting of any $n-2$ of $2n$ variables, then the corresponding $2n$-dimensional integral can be reduced to a one-dimensional oscillatory integral with, of course, a remainder. Hence, we need to give approximation expansions of one-dimensional oscillatory integrals. A basic expansion formula is shown as follows (see Section 4).

$$\begin{aligned} \int_0^1 f(x, \langle Nx \rangle dx &= \int_0^1 \int_0^1 f(x, y) dx dy \\ &+ \sum_{k=1}^{m} \frac{1}{k!} \left(\frac{1}{N} \right)^k \int_0^1 H_k(y, N) dy + O(N^{-m}). \end{aligned}$$

Here $f(x, y)$ is continuous on the square $[0, 1] \times [0, 1]$ and has a continuous mth partial derivative $f_x^{(m)} = (\partial/\partial x)^m f(x, y)$, and $H_k(y, N)$ is defined by

$$H_k(y, N) = f_x^{(k-1)}(1, y) \bar{B}_k(y - \langle N \rangle) - f_x^{(k-1)}(0, y) B_k(y),$$

where $B_k(x)$ and $\bar{B}_k(x)$ denote the *Bernoulli polynomial* of degree k and the corresponding *Bernoulli function* with period unity, respectively. The basic expansion formula shown above was proved in earlier literature and includes as its special cases some useful asymptotic expansions offered by Erugin-Sobolev, Krylov, Riekstenš, and Havie (see [26]). Detailed references, more special cases, and some expressions of the remainder term $O(N^{-m})$ can be found in Section 4. The DREs of reducing a $2n$-dimensional integral to a one-dimensional oscillatory integral with a remainder are shown in Section 3. As background knowledge, we describe the history of the problem in Section 1 and give the basic lemma in Section 2. Section 5 establishes some asymptotic expansion formulas for oscillatory integrals with singular factors.

3.1 DREs for approximating a double integral

A DRE can be established with the aid of an oscillatory integral. To describe this subject clearly, we probably need to start from the very beginning. In 1947, physicist A. Maréchal developed a mechanical quadrature to study the distribution of light (see [76] and [26]). We now repeat his idea as follows. Let $C : r = a\theta$, $a > 0$, be an Archimedean spiral in a circle of radius R, $f(r, \theta)$ a continuous function, $f(r, \theta + 2\pi) = f(r, \theta)$. Intuitively, if the positive number a is close enough to zero, then the line integral along C $2\pi a \int_C f(r, \theta)ds$ is reasonably close to the double integral over K, $\int_0^{2\pi} \int_0^R f(r, \theta)rdrd\theta$.

$$\lim_{a\to 0^+} 2\pi a \int_C f(r, \theta)ds = \int_0^{2\pi} \int_0^R f(r, \theta)rdrd\theta. \tag{3.1.1}$$

Equation (3.1.1) was not proven to be true until 1949, by Wilkins [76] (Grosswald [24] also proved it independently in 1951). We now give Wilkins' proof. From the differential formula for the arc measure $ds = (r^2 + (r'_\theta)^2)^{1/2}d\theta$ and $d\theta = dr/a$, we can rewrite equation (3.1.1) as

$$\lim_{a\to 0^+} 2\pi \int_0^R f(r, r/a)(a^2 + r^2)^{1/2}dr = \int_0^{2\pi} \int_0^R f(r, \theta)rdrd\theta.$$

Let M be $\max_{0\le r\le R, 0\le\theta\le 2\pi} f(r, \theta)$. Then

$$\begin{aligned} & |\int_0^R f(r, r/a)(a^2 + r^2)^{1/2}dr - \int_0^R f(r, r/a)rdr| \\ \le\; & M \int_0^R [(a^2 + r^2)^{1/2} - r]dr \to 0, \end{aligned}$$

as $a \to 0^+$. Hence, equation (3.1.1) is true if the following equation holds.

$$\lim_{a\to 0^+} 2\pi \int_0^R f(r, r/a)rdr = \int_0^{2\pi} \int_0^R f(r, \theta)rdrd\theta. \tag{3.1.2}$$

We now apply the Fejér theorem about Fourier summations. Denote

$$\begin{aligned} & S_n(r,\theta) \\ = \; & \tfrac{1}{2n\pi}\int_0^{\pi}\left[f(r,\theta+u)+f(r,\theta-u)\right]\tfrac{\sin^2(nu/2)}{\sin^2(u/2)}du \end{aligned}$$

and

$$A_{kn}(r) = \frac{1}{2\pi}\left(1-\frac{|k|}{n}\right)\int_0^{2\pi} f(r,\theta)(\cos k\theta + i\sin k\theta)d\theta.$$

Then, the Fejér polynomial can be written as

$$S_n(r,\theta) = \sum_{k=-(n-1)}^{n-1} A_{kn}(r)(\cos k\theta + i\sin k\theta).$$

From the Fejér theorem, $S_n(r,\theta)$ converges to $f(r,\theta)$ uniformly on K as $n\to\infty$. Therefore, for any $\epsilon > 0$, we can always find large enough n such that

$$|S_n(r,\theta) - f(r,\theta)| < \frac{\epsilon}{2\pi R^2}.$$

It follows that

$$\left|2\pi\int_0^R f(r,r/a)rdr - 2\pi\int_0^R S_n(r,r/a)rdr\right| < \frac{\epsilon}{2}. \tag{3.1.3}$$

On the other hand, we have

$$\begin{aligned} & \left|\int_0^R S_n(r,r/a)rdr - \int_0^R A_{0n}(r)rdr\right| \\ \le \; & \sum_{|k|=1}^{n-1}\left|\int_0^R A_{kn}(r)r\left(\cos\frac{kr}{a} + i\sin\frac{kr}{a}\right)dr\right|. \end{aligned}$$

From the Riemann–Lebesgue theorem, the right side of the above expression approaches zero as $a\to 0^+$. Hence, when $a > 0$ is small enough, we have

$$\left|2\pi\int_0^R S_n(r,r/a)rdr - 2\pi\int_0^R A_{0n}(r)rdr\right| < \frac{\epsilon}{2}. \tag{3.1.4}$$

The definition of $A_{0n}(r)$ gives

$$2\pi\int_0^R A_{0n}(r)rdr = \int_0^{2\pi}\int_0^R f(r,\theta)rdrd\theta.$$

Combining inequalities (3.1.3) and (3.1.4), we thus obtain

$$\left| 2\pi \int_0^R f(r, r/a) r dr - \int_0^{2\pi} \int_0^R f(r, \theta) r dr d\theta \right| < \epsilon,$$

which implies equation (3.1.1).

Put into words, equation (3.1.1) basically means that a double integral over a circular domain can be expressed as a limit of line integrals. Inspired by this idea, we think it is possible to develop a type of DRE such that a $2k$-dimensional integral can be expressed as limits of k-dimensional integrals. In the following section, we will show that this type of DRE exists and we will also find the approximation difference between the $2k$-dimensional integral and the k-dimensional integrals. All these results will contribute to the basic lemma in the next section.

3.2 Basic lemma

For the continuous function $f(x_1, x_2, \ldots, x_n)$, which is defined on the n-dimensional domain D, we denote its *modulus of continuity* by $\omega(f)$; i.e.,

$$\omega(f; \delta_1, \ldots, \delta_n) = \max \left| f(x_1, \ldots, x_n) - f(x_1', \ldots, x_n') \right|,$$

where $|x_i - x_i'| \leq \delta_i$, $i = 1, \ldots, n$. Let $V_n = [0, 1]^n$. In addition, for the sake of simplicity, we denote $X = (x_1, \ldots, x_k)$ and $\langle NX \rangle = (\langle N_1 x_1 \rangle, \ldots, \langle N_k x_k \rangle)$, where $\langle x \rangle$ is the fractional part of the nonnegative real number x; i.e., $\langle x \rangle = x - [x]$. Finally, we let $C(V_n \times V_n)$ be the set of all continuous functions defined on $V_n \times V_n$. For $f(x_1, \ldots, x_n; y_1, \ldots, y_n) \in C(V_n \times V_n)$, we now study the approximation to the integral

$$I = \int_0^1 \cdots \int_0^1 f(x_1, \ldots, x_n; y_1, \ldots, y_n) dx_1 \cdots dx_n dy_1 \cdots dy_n.$$

We will approximate integral I by using

$$\begin{aligned} & I_{N_1 \cdots N_n}(f) \\ = \; & \textstyle\int_0^1 \cdots \int_0^1 f(x_1, \ldots, x_n; \langle N_1 x_1 \rangle, \cdots, \langle N_n y_n \rangle) dx_1 \cdots dx_n \end{aligned}$$

and denoting the error by

$$\rho_{N_1 \cdots N_n}(f) = I_{N_1 \cdots N_n}(f) - I. \tag{3.2.1}$$

Lemma 3.2.1 *For any $f \in C$ and positive integers $N_i \geq 2$, $i = 1, \ldots, n$, there exists*

$$\begin{aligned} & \left| \rho_{N_1 \cdots N_n}(f) \right| \\ \leq \; & 2^n \int_0^1 \cdots \int_0^1 \omega\left(f; \frac{t_1}{N_1}, \ldots, \frac{t_n}{N_n}\right) \Pi_{j=1}^n (1 - t_j) dt_1 \cdots dt_n. \end{aligned}$$

In addition, there exists a nontrivial 2n-variable function $f_0 \in C$ for which the equality in (3.2.1) is satisfied.

Proof. By using a suitable transform, we have

$$\begin{aligned}
&\rho_{N_1\cdots N_n}(f)\\
= &\left(\Pi_{j=1}^n N_j\right)^{-1}\int_0^{N_1}\cdots\int_0^{N_n} f\left(\frac{y_1}{N_1},\dots,\frac{y_n}{N_n};\langle y_1\rangle,\dots,\langle y_n\rangle\right)dy_1\cdots dy_n\\
&-\int_0^1\cdots\int_0^1\left[\sum_{k_1=0}^{N_1-1}\cdots\sum_{k_n=0}^{N_n-1}\int_{\frac{k_1}{N_1}}^{\frac{k_1+1}{N_1}}\cdots\int_{\frac{k_n}{N_n}}^{\frac{k_n+1}{N_n}} f(x_1,\dots,x_n;y_1,\dots,y_n)\right.\\
&\left.\times dx_1\cdots dx_n\right]dy_1\cdots dy_n\\
= &\left(\Pi_{j=1}^n N_j\right)^{-1}\sum_{k_1=0}^{N_1-1}\cdots\sum_{k_n=0}^{N_n-1}\int_0^1\cdots\int_0^1\left[f\left(\frac{y_1+k_1}{N_1},\dots,\frac{y_n+k_n}{N_n};\right.\right.\\
&\left. y_1,\dots,y_n\right)-\int_0^1\cdots\int_0^1 f\left(\frac{t_1+k_1}{N_1},\dots,\frac{t_n+k_n}{N_n};y_1,\dots,y_n\right)\\
&\left.\times dt_1\cdots dt_n\right]dy_1\cdots dy_n. \qquad (3.2.2)
\end{aligned}$$

Therefore,

$$\begin{aligned}
&\left|\rho_{N_1\cdots N_n}(f)\right|\\
\le\ &2^n\int_0^1\cdots\int_0^1\omega\left(f;\tfrac{t_1}{N_1},\dots,\tfrac{t_n}{N_n}\right)\Pi_{j=1}^n(1-t_j)dt_1\cdots dt_n.
\end{aligned}$$

Now, we construct a function, f_0, for which the equality in (3.2.1) is satisfied. Denote

$$g_{k,j}(x,y)=x+\frac{2k}{N_j}(y-1)$$

and

$$h_{k,j}(x,y)=-x+\frac{2k+2}{N_j}y.$$

Define

$$\begin{aligned}
&\phi_j(x,y)\\
= &\begin{cases} g_{k,j}(x,y), & \text{if}\frac{k}{N_j}\le x\le\frac{k+1}{N_j}, N_jx-k\le y\le 1,\\ h_{k,j}(x,y), & \text{if}\frac{k}{N_j}\le x\le\frac{k+1}{N_j}, 0\le y\le N_jx-k,\end{cases}
\end{aligned}$$

where $k=0,1,\dots,N_j-1$. Let

$$f_0(x_1,\dots,x_n;y_1,\dots,y_n)=\sum_{j=1}^n\phi_j(x_j,y_j).$$

Then $f_0 \in C$, and $\omega(f_0; t_1, \ldots, t_n) = t_1 + \cdots + t_n$. Substituting f_0 into the right side of (3.2.1), we have

$$2^n \int_0^1 \cdots \int_0^1 \omega\left(f_0; \frac{t_1}{N_1}, \ldots, \frac{t_n}{N_n}\right) \Pi_{j=1}^n (1-t_j)dt_1 \cdots dt_n$$
$$= \frac{1}{3} \sum_{j=1}^n \frac{1}{N_j}.$$

Using equation (3.2.2),

$$\rho_{N_1 \cdots N_n}(f_0) = \frac{1}{3} \sum_{j=1}^n \frac{1}{N_j}.$$

Hence, the equality in (3.2.1) holds for f_0. □

Lemma 3.2.2 *For all positive integers $N_1, \ldots, N_n$,*

$$\delta_{N_1 \cdots N_n} = \sup_{\omega\left(f; \frac{1}{N_1}, \ldots, \frac{1}{N_n}\right) \neq 0} \frac{|\rho_{N_1 \ldots N_n}(f)|}{\omega\left(f; \frac{1}{N_1}, \ldots, \frac{1}{N_n}\right)} = 1.$$

Proof. From expression (3.2.1), we immediately have $\delta_{N_1 \cdots N_n} \leq 1$. Let $g_j(x)$ be a continuous periodic function defined on $(-\infty, \infty)$ with period $1/N_j$ and of the expression

$$g_j(x) = \begin{cases} x^\alpha, & \text{if } 0 \leq x \leq \frac{1}{2N_j}, \\ \left(\frac{1}{N_j} - x\right)^\alpha, & \text{if } \frac{1}{2N_j} \leq x \leq \frac{1}{N_j}, \end{cases}$$

where $x \in [0, 1/N_j]$ and $0 < \alpha < 1$. Define the function

$$f_0(x_1, \ldots, x_n; y_1, \ldots, y_n) = \sum_{j=1}^n g_j\left(x_j - \frac{y_j}{N_j}\right).$$

Then $f_0 \in C$ and satisfies

$$\frac{|\rho_{N_1 \cdots N_n}(f_0)|}{\omega\left(f_0; \frac{1}{N_1}, \ldots, \frac{1}{N_n}\right)} = \frac{\frac{1}{(\alpha+1)2^\alpha} \sum_{j=1}^n \frac{1}{N_j^\alpha}}{\sum_{j=1}^n \frac{1}{(2N_j)^\alpha}} = \frac{1}{\alpha+1}.$$

Since α can be arbitrarily close to zero , $\delta_{N_1 \cdots N_n} \geq 1$. This completes the proof of Lemma 3.2.2. □

From Lemmas 3.2.1 and 3.2.2, we immediately obtain the following basic lemma.

Lemma 3.2.3 (Basic Lemma) *For any $f \in C$ and positive integers $N_i \geq 2$ $(i = 1, \dots, n)$, $\rho_{N_1 \cdots N_n}(f)$ defined by (3.2.1) satisfies*

$$|\rho_{N_1 \cdots N_n}(f)| \leq \omega\left(f; \frac{1}{N_1}, \dots, \frac{1}{N_n}\right). \tag{3.2.3}$$

Given a continuous modular function $\omega(t_1, \dots, t_n)$, denote all functions in C that satisfy $\omega(f; t_1, \dots, t_n) \leq \omega(t_1, \dots, t_n)$ by H^ω. In particular, if $\omega(t_1, \dots, t_n) = t_1 + \cdots + t_n$, H^ω is simply denoted by H'. We now give the properties of

$$E_{N_1 \cdots N_n}(H^\omega) = \sup_{f \in H^\omega} |\rho_{N_1 \cdots N_n}(f)|.$$

First, from Lemma 3.2.1, we have

$$E_{N_1 \cdots N_n}(H') = \frac{1}{3} \sum_{j=1}^{n} \frac{1}{N_j}.$$

For the general case, we give the following result.

Theorem 3.2.4 *Let $N_1, \dots, N_n$ be positive integers. For a given continuous module $\omega(t_1, \dots, t_n)$, the asymptotic formula*

$$\begin{aligned} E_{N_1 \cdots N_n}(H^\omega) &= 2^n \int_0^{\frac{1}{2}} \cdots \int_0^{\frac{1}{2}} \omega\left(\frac{t_1}{N_1}, \dots, \frac{t_n}{N_n}\right) dt_1 \dots dt_n \\ &\quad + O\left(\frac{1}{N_1} + \cdots + \frac{1}{N_n}\right) \end{aligned} \tag{3.2.4}$$

holds, in which O depends on ω only.

Proof. Let $f \in H^\omega$, then

$$\begin{aligned} &\rho_{N_1 \cdots N_n}(f) \\ = \ & \int_0^1 \cdots \int_0^1 \Bigg[\left(\Pi_{j=1}^n N_j\right)^{-1} \sum_{k_1=1}^{N_1-2} \cdots \sum_{k_n=1}^{N_n-2} \\ & \times f\left(\tfrac{y_1+k_1}{N_1}, \dots, \tfrac{y_n+k_n}{N_n}; y_1, \dots, y_n\right) - \sum_{k_1=1}^{N_1-2} \cdots \sum_{k_n=1}^{N_n-2} \\ & \times \int_{\frac{y_1+k_1-\frac{1}{2}}{N_1}}^{\frac{y_1+k_1+\frac{1}{2}}{N_1}} \cdots \int_{\frac{y_n+k_n-\frac{1}{2}}{N_n}}^{\frac{y_n+k_n+\frac{1}{2}}{N_n}} f(x_1, \dots, x_n; y_1, \dots, y_n)\, dx_1 \cdots dx_n \Bigg] \\ & \times dy_1 \cdots dy_n + O\left(\tfrac{1}{N_1} + \cdots + \tfrac{1}{N_n}\right). \end{aligned}$$

Obviously, in the above expression, O depends on ω only. We thus obtain

$$\begin{aligned} &|\rho_{N_1 \cdots N_n}(f)| \\ \leq \ & 2^n \int_0^{\frac{1}{2}} \cdots \int_0^{\frac{1}{2}} \omega\left(\tfrac{t_1}{N_1}, \dots, \tfrac{t_n}{N_n}\right) dt_1 \cdots dt_n + O\left(\tfrac{1}{N_1} + \cdots + \tfrac{1}{N_n}\right). \end{aligned}$$

Therefore, the left-hand side of equation (3.2.4) is less than or equal to the right-hand side of equation (3.2.4).

We now prove that the left-hand side of equation (3.2.4) is no less than the right-hand side. Let $g(x_1, \ldots, x_n)$ be an n-variable periodic continuous function that has period $\frac{1}{N_j}$ for variable x_j and is defined on $\left[0, \frac{1}{N_1}\right] \times \left[0, \frac{1}{N_2}\right] \times \cdots \left[0, \frac{1}{N_n}\right]$ by

$$g(\cdots, x_j, \cdots) = \begin{cases} \omega(\ldots, x_j, \ldots), & x_j \in \left[0, \frac{1}{2N_j}\right], \\ \omega\left(\ldots, \frac{1}{N_j} - x_j, \ldots\right), & x_j \in \left[\frac{1}{2N_j}, \frac{1}{N_j}\right], \end{cases}$$

for $j = 1, \ldots, n$. Let

$$f_0(x_1, \ldots, x_n; y_1, \ldots, y_n) = g\left(x_1 - \frac{y_1}{N_1}, \ldots, x_n - \frac{y_n}{N_n}\right).$$

Then $f_0 \in C$. It is not hard to see that $f_0 \in H^\omega$ and

$$|\rho_{N_1 \cdots N_n}(f_0)| = 2^n \int_0^{\frac{1}{2}} \cdots \int_0^{\frac{1}{2}} \omega\left(\frac{t_1}{N_1}, \ldots, \frac{t_n}{N_n}\right) dt_1 \cdots dt_n.$$

Therefore, in equation (3.2.4), its left-hand side is no less than its right-hand side, thus completing the proof of Theorem 3.2.4. □

Corollary 3.2.5 *Let $f(x, y)$ be a continuous function defined on $0 \le x, y \le 1$. Then*

$$\lim_{N \to \infty} \int_0^1 f(x, \langle Nx \rangle) dx = \int_0^1 \int_0^1 f(x, y) dx dy, \tag{3.2.5}$$

where $\langle x \rangle$, as we showed at the beginning of the section, is the fractional part of the nonnegative real number x; i.e., $\langle x \rangle = x - [x]$.

Corollary 3.2.6 *Assume $f(x, y)$ satisfies the* Lipschitz condition

$$|f(x, y) - f(x', y)| \le A|x - x'|, \qquad 0 \le x \le x' \le 1,$$

where A is a constant that is independent of $y (0 \le y \le 1)$. Then for all integers $N \ge 2$, we have

$$\left| \int_0^1 \int_0^1 f(x, y) dx dy - \int_0^1 f(x, \langle Nx \rangle) dx \right| \le \frac{A}{N}. \tag{3.2.6}$$

We should point out that the estimate $O\left(\frac{1}{N}\right)$ on the right-hand side of the above equation cannot be improved. In fact, if $f(x, y) = xy$, then

$$\left| \int_0^1 \int_0^1 xy dx dy - \int_0^1 x \langle Nx \rangle dx \right| = \frac{1}{12N}$$

for $N = 2, 3, \ldots$, which is exactly of order N^{-1}.

Corollary 3.2.7 *Let $g(r,\theta)$ be a continuous function defined on the circular domain $S(0 \le r \le r_0, 0 \le \theta \le 2\pi)$ and let $g(r, 2\pi+\theta) = g(r,\theta)$. Denote*

$$\omega_g(\delta,\delta') = \max_S |g(r,\theta) - g(r',\theta')|, \quad |r-r'| \le \delta, \ |\theta-\theta'| \le \delta'.$$

Then, for all integers $N \ge 2$,

$$\begin{aligned} &\left| \iint_S g(r,\theta)dS - 2\pi \int_0^{r_0} g\left(r, \frac{2N\pi r}{r_0}\right) r dr \right| \\ \le \ & 2\pi r_0^2 \left[(r_0+1)\omega\left(g; \frac{1}{N}, 0\right) + M_g \frac{1}{N} \right], \end{aligned} \tag{3.2.7}$$

where $M_g = \max_S |g(r,\theta)|$.

Proof. First, take the transform

$$\tilde{r} = \frac{r}{r_0}, \quad \tilde{\theta} = \frac{\theta}{2\pi}, \quad \theta(r_0\tilde{r}, 2\pi\tilde{\theta})\tilde{r} = f(\tilde{r},\tilde{\theta}).$$

$f(\tilde{r},\tilde{\theta})$ has period 1 with respect to $\tilde{\theta}$. We thus have

$$\iint_S g(r,\theta)dS = 2\pi r_0^2 \int_0^1 \int_0^1 f(\tilde{r},\tilde{\theta}) d\tilde{r} d\tilde{\theta} \tag{3.2.8}$$

and

$$\begin{aligned} 2\pi \int_0^{r_0} g\left(r, 2N\pi \frac{r}{r_0}\right) r dr &= 2\pi r_0^2 \int_0^1 f(\tilde{r}, N\tilde{r}) d\tilde{r} \\ &= 2\pi r_0^2 \int_0^1 f(\tilde{r}, \langle N\tilde{r}\rangle) d\tilde{r}. \end{aligned} \tag{3.2.9}$$

Denote the domain $0 \le x \le 1, 0 \le y \le 2\pi$ by Δ. We have then, for x and x' that satisfy $|x-x'| \le 1/N$,

$$\begin{aligned} & \omega\left(f; \tfrac{1}{N}, 0\right) \\ = \ & \max_\Delta \left| g(r_0x, y)x - g(r_0x', y)x' \right| \\ \le \ & \max_\Delta \left| g(r_0x, y)x - g(r_0x', y)x \right| + \max_\Delta \left| g(r_0x', y) \right| |x-x'| \\ \le \ & \omega\left(g; \tfrac{r_0}{N}, 0\right) + M_g|x-x'| \\ \le \ & (r_0+1)\omega\left(g; \tfrac{1}{N}, 0\right) + M_g \tfrac{1}{N}. \end{aligned}$$

Hence, from expressions (3.2.8) and (3.2.9) and using the basic lemma (Lemma 3.2.3), we obtain asymptotic formula (3.2.7). □

Comparing expressions (3.1.2) and (3.2.7), we know that Corollary 3.2.7 improved the Maréchal–Wilkins formula (3.1.2).

From the proof of the basic lemma, it is not hard to obtain the following generalization of Corollary 3.2.5 by using a linear transform for $n = 1$.

$$\lim_{N\to\infty} \int_{\alpha}^{\beta} f(x, \langle Nx \rangle) dx = \int_{\alpha}^{\beta} dx \int_{0}^{1} f(x, y) dy, \tag{3.2.10}$$

where $f(x, y)$ is assumed to be a continuous function defined on $\alpha < x < \beta$, $0 \le y \le 1$.

Making use of equation (3.2.10), we obtain the following result.

Corollary 3.2.8 *Let V_3 be a 3-dimensional cubic $[0, 1]^3$ and $f(x, y, z) \in C(V_3)$. For positive integers λ and μ,*

$$\lim_{\mu\to\infty} \lim_{\lambda\to\infty} \int_{0}^{1} f(x, \langle \lambda x \rangle, \langle \mu x \rangle) dx = \int_{0}^{1} \int_{0}^{1} \int_{0}^{1} f(x, y, z) dx dy dz. \tag{3.2.11}$$

Proof. In the basic lemma take $k = 2$ and $N_1 = N_2 = \mu$, then from (3.2.3),

$$\int_{0}^{1} \int_{0}^{1} \int_{0}^{1} \int_{0}^{1} f(x, y, z) dx dy dz du = \lim_{\mu\to\infty} \int_{0}^{1} \int_{0}^{1} f(x, y, \langle \mu x \rangle) dx dy.$$

We know that $f(x, y, \langle \mu x \rangle)$ is a bounded function and is continuous on the rectangle domain $\frac{\nu-1}{\mu} < x < \frac{\nu}{\mu}$, $0 \le y \le 1$ $(\nu = 1, 2, \ldots, \mu)$. Therefore, by using equation (3.2.10), we have

$$\begin{aligned}
\int_0^1 \int_0^1 f(x, y, \langle \mu x \rangle) dx dy &= \sum_{\nu=1}^{\mu} \int_{(\nu-1)/\mu}^{\nu/\mu} dx \int_{0}^{1} f(x, y, \langle \mu x \rangle) dy \\
&= \sum_{\nu=1}^{\mu} \lim_{\lambda\to\infty} \int_{(\nu-1)/\mu}^{\nu/\mu} f(x, \langle \lambda x \rangle, \langle \mu x \rangle) dx \\
&= \lim_{\lambda\to\infty} \int_{0}^{1} f(x, \langle \lambda x \rangle, \langle \mu x \rangle) dx,
\end{aligned}$$

completing the proof of equation (3.2.11). □

3.3 DREs with large parameters

An integral over $V_{n-1} = [0, 1]^{n-1}$ can always be expressed as an integral over $V_n = [0, 1]^n$; i.e.,

$$\begin{aligned}
&\int_{V_{n-1}} f(x_1, \ldots, x_{2n-1}) dx_1 \cdots dx_{2n-1} \\
= &\int_{V_n} f(x_1, \ldots, x_{2n-1}) dx_1 \cdots dx_{2n-1} dx_{2n},
\end{aligned}$$

where $\int_{V_{n-1}}$ and $\int_{V_n}$ are $(n-1)$-dimensional and n-dimensional integrals, respectively. Therefore, from the basic lemma, we can approximate an n-dimensional integral over V_n by an $\left[\frac{n+1}{2}\right]$-dimensional integral. The approximation error can be described by the modulus of the continuity of the integrand.

For the purpose of simplicity, we assume that in the new integral of $\left[\frac{n+1}{2}\right]$ variables, k large parameters $N_1, \dots, N_k$ are the same. Hence, after each integral dimension reducing, only one parameter will be introduced. In addition, an n-dimensional integral will be reduced into a single integral by dimension reducing at most $[\log_2 n] + 1$ times. We will give more details as follows.

Let $F(x_1, \dots, x_n)(n \geq 3)$ be a continuous function defined on V_n, with each of its variables $x_3, \dots, x_n$ having a period of 1 (i.e., $F(x_1, x_2, \dots, x_i + 1, \dots, x_n) = F(x_1, x_2, \dots, x_i, \dots, x_n)$, $i = 3, \dots, n$). Also let s be an integer that satisfies $2^{s-1} < n \leq 2^s$. After dimension reducing s times, the n-dimensional integral $\int_{V_n} F dx_1 \cdots dx_n$ can be reduced to a single integral $\int_0^1 \psi(x_1)dx_1$ with a remainder term $\rho(N_1, \dots, N_s)$. Here, $\rho(N_1, \dots, N_s) \to 0$ $(N_i \to \infty)$, and $\psi(x_1)$ is a piecewise continuous function in x_1 with s large parameters $N_1, \dots, N_s$. More precisely, we can obtain

$$\begin{aligned} & \int_{V_n} F(x_1, \dots, x_n) dx_1 \cdots dx_n \\ = & \int_0^1 \psi(x_1) dx_1 + \rho(N_1, \dots, N_s), \end{aligned}$$

where $\psi(x_1) = F(x_1, y_1, \dots, y_{n-1})$ and $(y_1, \dots, y_{n-1})$ is a permutation of $n-1$ elements chosen from the following $2^s - 1$ elements

$$\langle N_{\nu_1} \langle N_{\nu_2} \langle \cdots \langle N_{\nu_t} x_1 \rangle \cdots \rangle\rangle\rangle,$$

for $1 \leq \nu_1 < \nu_2 < \cdots < \nu_t \leq s$ and $1 \leq t \leq s$.

It is easy to understand that the periodicity of function F with respect to $x_3, \dots, x_n$ insures the piecewise continuity of $\psi(x_1) = F(x_1, y_1, \dots, y_{n-1})$ on the interval $0 \leq x_1 \leq 1$. We use as an example the case of $n = 4$ to demonstrate it. Assume that the continuous function $f(x_1, x_2, x_3, x_4)$ defined on V_4 satisfies $f(x_1, x_2, 0, x_4) = f(x_1, x_2, 1, x_4)$ and $f(x_1, x_2, x_3, 0) = f(x_1, x_2, x_3, 1)$. Then, $\phi(x_1, x_2) = f(x_1, x_2, \langle \lambda x_1 \rangle, \langle \mu x_2 \rangle)$ is clearly a continuous function. It follows that $\psi(x_1) = \phi(x_1, \langle \nu x_1 \rangle) = f(x_1, \langle \nu x_1 \rangle, \langle \lambda x_1 \rangle, \langle \mu \langle \nu x \rangle \rangle)$ is at least a piecewise continuous function of x_1. Here, λ, μ, and ν can be any integers. Moreover, if $f(x_1, 0, x_3, x_4) = f(x_1, 1, x_3, x_4)$, then $\psi(x_1)$ is continuous.

From the above analysis, in order to reduce an n-dimensional integral eventually into a single integral, we need the integrand to have the period 1 for its $n-2$ variables. In fact, if the n-dimensional integral is reduced once by using the basic lemma, the lower dimensional integral is generally piecewise continuous; it may or may not be continuous at $x_i = \nu / N_i$, $1 \leq \nu \leq N_i - 1$. Therefore, in general, the above dimensionality reducing process is only suitable for multivariate integration of continuous multivariate periodic functions. However, if the integral domain is an n-dimensional sphere, then the requirement on the periodicity of the integrand in terms of its $n-2$ variables can be released. In other words, our dimensionality reducing process is more suitable for hyperspheres.

We also need to point out that, in order to decrease the approximation error, we should choose the large parameters, $N_1, \ldots, N_n$, carefully when applying the basic lemma to reduce a $2n$-dimensional integral into an n-dimensional integral. In general, to evaluate the n-dimensional integral $\int_{V_n} f(x_1, \ldots, x_n; \langle N_1 x_1 \rangle, \ldots, \langle N_n x_n \rangle) dx_1 \cdots, dx_n$ easily, we need the parameters $N_1, \ldots, N_n$ chosen to be not too large. But, on the other hand, we need the size of the range of the approximation errors,

$$2\omega\left(f; \frac{1}{N_1}, \ldots, \frac{1}{N_n}, 0, \ldots, 0\right),$$

to be the smallest possible. Therefore, we should consider the rate of change of $f(x_1, \ldots, x_n)$ along each x_i-axis. If, for instance, f has a large rate of change along the x_i-axis, then a large corresponding parameter N_i will be chosen.

In summary, we have two principles for the dimensionality-reduction shown above: the periodicity of the integrand and a suitable choice of parameters. In the following, the dimension reducing process and the principles will be used to find the DREs for some lower dimensional integrals.

Theorem 3.3.1 *Let $f(x, y, z)$ be a continuous function defined on $V_3 = [0, 1]^3$, that satisfies $f(x, y, 0) = f(x, y, 1)$. Then,*

$$\int_0^1 \int_0^1 \int_0^1 f dx dy dz = \int_0^1 f(x, \langle \mu x \rangle, \langle \lambda x \rangle) dx + \rho(\lambda, \mu),$$

where $|\rho(\lambda, \mu)| \leq \omega\left(\phi; \frac{1}{\mu}, 0\right) + \omega\left(f; \frac{1}{\lambda}, 0, 0\right)$ and $\phi = \phi_\lambda = f(x, y, \langle \lambda x \rangle)$ is a continuous function with respect to (x, y). Here, $\lambda, \mu \geq 2$ are integers.

Proof. Obviously, $f(x, y, \langle \lambda x \rangle)$ is continuous in terms of (x, y), and $f(x, \langle \mu x \rangle, \langle \lambda x \rangle)$ is a piecewise continuous function in terms of x. We now rewrite the 3-dimensional integral on the left-hand side of the above equation as a 4-dimensional integral

$$\int_0^1 \int_0^1 \int_0^1 \int_0^1 f(x, y, z) dx dy dz du,$$

and apply the basic lemma twice. Then the result of the theorem is found. □

Because

$$\lim_{\mu \to \infty} \omega\left(\phi; \frac{1}{\mu}, 0\right) = 0, \quad \lim_{\lambda \to \infty} \omega\left(f; \frac{1}{\lambda}, 0, 0\right) = 0,$$

we have $\lim_{\lambda\to\infty}\lim_{\mu\to\infty}\rho(\lambda,\mu)=0$. Therefore, for large enough λ and μ, we obtain

$$\int_0^1\int_0^1\int_0^1 f(x,y,z)dxdydz \approx \int_0^1 f(x,\langle\mu x\rangle,\langle\lambda x\rangle)dx.$$

The approximation is the DRE for the integral $\int_0^1\int_0^1\int_0^1 f(x,y,z)dx\,dydz$ with large parameters μ and λ.

Theorem 3.3.2 *Let $f(x,y,z)$ be a continuous function defined on $V_3=[0,1]^3$ that satisfies $f(x,y,0)=f(x,y,1)$. Then, for all integers $N\geq 2$*

$$\begin{aligned}&\left|\int_0^1\int_0^1\int_0^1 f(x,y,z)dxdydz - \int_0^1 f(x,\langle N^2x\rangle,\langle Nx\rangle)dx\right|\\ \leq\ & 4\omega\left(f;\frac{1}{N},0,\frac{1}{N}\right).\end{aligned}$$

Proof. First of all, from Theorem 3.3.1,

$$\begin{aligned}&\left|\int_0^1\int_0^1\int_0^1 f(x,y,z)dxdydz - \int_0^1 f(x,\langle N^2x\rangle,\langle Nx\rangle)dx\right|\\ \leq\ & \omega\left(f;\frac{1}{N},0,0\right)+\omega\left(\phi;\frac{1}{N^2},0\right),\end{aligned}\tag{3.3.1}$$

where $\phi=\phi_N=f(x,y,\langle Nx\rangle)$. Now let us estimate $\omega(\phi;N^{-2},0)$. Denote $I_k=\left[\frac{k-1}{N},\frac{k}{N}\right]$, $k=1,2,\ldots,N$. For any $x,x'\in I_k$ with $|x-x'|\leq N^{-2}$, $|\langle Nx\rangle-\langle Nx'\rangle|\leq N^{-1}$. It follows that

$$\begin{aligned}&\left|f(x,y,\langle Nx\rangle)-f(x',y,\langle Nx'\rangle)\right|\\ \leq\ & \omega\left(f;\tfrac{1}{N^2},0,\tfrac{1}{N}\right).\end{aligned}$$

In other cases of $x,x'\in[0,1]$ and $|x-x'|\leq N^{-2}$. We can always assume that $x\in I_{k+1}$ and $x'\in I_k$. Thus, x and x' can be written as $x=\frac{k}{N}+\epsilon_1$ and $x'=\frac{k}{N}-\epsilon_2$, where $\epsilon_1,\epsilon_2\geq 0$ and $\epsilon_1+\epsilon_2\leq N^{-2}$. Therefore, we have

$$\begin{aligned}&\left|f(x,y,\langle Nx\rangle)-f(x',y,\langle Nx'\rangle)\right|\\ \leq\ & \left|f(x,y,\langle Nx\rangle)-f(x',y,\langle Nx\rangle)\right|\\ &+\left|f(x',y,\langle Nx\rangle)-f(x',y,\langle Nx'\rangle)\right|\\ \leq\ & \omega\left(f;\tfrac{1}{N^2},0,0\right)+\left|f(x',y,\langle N\epsilon_1\rangle)-f(x',y,\langle 1-N\epsilon_2\rangle)\right|\\ \leq\ & \omega\left(f;\tfrac{1}{N^2},0,0\right)+\left|f(x',y,\langle N\epsilon_1\rangle)-f(x',y,0)\right|\\ &+\left|f(x',y,1)-f(x',y,\langle 1-N\epsilon_2\rangle)\right|\\ \leq\ & \omega\left(f;\tfrac{1}{N^2},0,0\right)+2\omega\left(f;0,0,\tfrac{1}{N}\right)\\ \leq\ & 3\omega\left(f;\tfrac{1}{N},0,\tfrac{1}{N}\right).\end{aligned}$$

Hence, $\omega(\phi;N^{-2},0)\leq 3\omega(f;N^{-1},0,N^{-1})$. Substituting this estimate into the right-hand side of expression (3.3.1), we complete the proof of the theorem. □

Obviously, Theorem 3.3.2 is better for application than Theorem 3.3.1 because the former contains only one large parameter and has a simple error estimation $8\omega(f; N^{-1}, 0, N^{-1})$.

Consider the 3-dimensional integral

$$J = \int_\sigma F(r, \phi, \theta) d\sigma,$$

where (r, ϕ, θ) are the spherical coordinates, $F(r, \phi, \theta)$ is a continuous function defined on the sphere $\sigma(0 \le r \le R, 0 \le \phi \le \pi, 0 \le \theta \le 2\pi)$, and $d\sigma$ is the spherical surface measure. By using the relations $F(r, \phi, \theta) = F(r, 2\pi - \phi, \theta)$ and $F(r, \phi, \theta) = F(r, \phi + 2k\pi, \theta) = F(r, \phi, \theta + 2k\pi)$, $k = 1, 2, 3$, we can extend the domain of the continuous function $F(r, \phi, \theta)$ to $\sigma^*(0 \le r \le R, 0 \le \phi \le \infty, 0 \le \theta \le \infty)$; i.e., F becomes a 2π periodic function with respect to ϕ and θ, respectively, and keeps the continuity over the entire domain σ^*. Thus,

$$\begin{aligned} J &= \tfrac{1}{2} \int_0^{2\pi} d\phi \int_0^{2\pi} d\theta \int_0^R F(r, \phi, \theta) r^2 |\sin\phi| dr \\ &= 2\pi^2 R^3 \int_0^1 \int_0^1 \int_0^1 f(x, y, z) dx dy dz, \end{aligned}$$

where $f(x, y, z) = F(Rx, 2\pi z, 2\pi y) x^2 |\sin(2\pi z)|$ and $f(x, y, z)$ possesses a periodicity of 1 with respect to both y and z.

In order to apply Theorem 3.3.2 to J, we need to estimate $\omega(f; N^{-1}, 0, N^{-1})$. Let x and x' satisfy $|x - x'| \le N^{-1}$, and z, and z' satisfy $|z - z'| \le N^{-1}$. Then

$$\begin{aligned} &\omega(f; N^{-1}, 0, N^{-1}) \\ = \;& \max \big| F(Rx, 2\pi z, 2\pi y) x^2 |\sin(2\pi z)| \\ & - F(Rx', 2\pi z', 2\pi y)(x')^2 |\sin(2\pi z')| \big| \\ \le \;& \omega\left(F; \tfrac{R}{N}, \tfrac{2\pi}{N}, 0\right) |x^2 \sin(2\pi z)| \\ & + C_F |x^2 \sin(2\pi z) - (x')^2 \sin(2\pi z')| \\ \le \;& \omega\left(F; \tfrac{R}{N}, \tfrac{2\pi}{N}, 0\right) + C_F |x^2 - (x')^2| |\sin(2\pi z)| \\ & + C_F |x'|^2 |\sin(2\pi z) - \sin(2\pi z')| \\ \le \;& \omega\left(F; \tfrac{R}{N}, \tfrac{2\pi}{N}, 0\right) + 2C_F \tfrac{1}{N} \\ & + 2C_F |\sin\pi(z - z') \cos\pi(z + z')| \\ \le \;& \omega\left(F; \tfrac{R}{N}, \tfrac{2\pi}{N}, 0\right) + 2C_F \tfrac{1}{N} + 2C_F \left|\sin\tfrac{\pi}{N}\right| \\ \le \;& (R + 1)(2\pi + 1) \omega\left(F; \tfrac{1}{N}, \tfrac{1}{N}, 0\right) + (2 + 2\pi) C_F \tfrac{1}{N}, \end{aligned}$$

where $N \ge 2$ is an integer and C_F is a constant that depends only on the function F. Similarly, if we define $g(x, y, z) = F(Rx, 2\pi y, 2\pi z) x^2 |\sin(2\pi y)|$ and $h(x, y, z) = F(Ry, 2\pi x, 2\pi z) y^2 |\sin(2\pi x)|$, then the corresponding error estimates are

$$\begin{aligned} &\omega(g; N^{-1}, 0, N^{-1}) \\ \le \;& (R + 1)(2\pi + 1) \omega\left(F; \tfrac{1}{N}, 0, \tfrac{1}{N}\right) + 2C_F \tfrac{1}{N} \end{aligned}$$

and

$$\begin{aligned}&\omega(h; N^{-1}, 0, N^{-1})\\ \leq\ & (2\pi+1)^2\omega\left(F; 0, \tfrac{1}{N}, \tfrac{1}{N}\right) + 2\pi C_F \tfrac{1}{N},\end{aligned}$$

respectively.

Now we apply Theorem 3.3.2 to the integral J. Then, we have

Theorem 3.3.3 *Let $F(r, \phi, \theta)$ be a continuous function defined on sphere $\sigma\,(0 \leq r \leq R)$, which satisfies $F(r, \phi, \theta) = F(r, 2\pi - \phi, \theta)$ and has period 2π with respect to both ϕ and θ. Then for all integers $N \geq 2$, we have*

$$\begin{aligned}&\Big|\tfrac{1}{2\pi^2 R^3}\textstyle\int_\sigma F(r, \phi, \theta) d\sigma - \int_0^1 F(Rx, 2\pi N x, 2\pi N^2 x)\\ &\quad \times x^2 |\sin(2\pi N x)| dx\Big|\\ \leq\ & 4\left((R+1)(2\pi+1)\omega(F; N^{-1}, N^{-1}, 0) + 2(\pi+1) C_F N^{-1}\right).\end{aligned}$$

Similarly, the following two results hold.

Theorem 3.3.4 *Under the same assumptions as Theorem 3.3.3, we have*

$$\begin{aligned}&\Big|\tfrac{1}{2\pi^2 R^3}\textstyle\int_\sigma F(r, \phi, \theta) d\sigma - \int_0^1 F(Rx, 2\pi N^2 x, 2\pi N x)\\ &\quad \times x^2 |\sin(2\pi N^2 x)| dx\Big|\\ \leq\ & 4\left((R+1)(2\pi+1)\omega(F; N^{-1}, 0, N^{-1}) + 2 C_F N^{-1}\right).\end{aligned}$$

Theorem 3.3.5 *Under the same assumptions as Theorem 3.3.3, we have*

$$\begin{aligned}&\Big|\tfrac{1}{2\pi^2 R^3}\textstyle\int_\sigma F(r, \phi, \theta) d\sigma - \int_0^1 F(R\langle N^2 x\rangle, 2\pi x, 2\pi N x)\\ &\quad \times (\langle N^2 x\rangle)^2 |\sin(2\pi x)| dx\Big|\\ \leq\ & 4\left((2\pi+1)^2\omega(F; 0, N^{-1}, N^{-1}) + 2\pi C_F N^{-1}\right).\end{aligned}$$

Theorems 3.3.3, 3.3.4, and 3.3.5 present ways to approximate a triple integral over a sphere by using limits of line integrals with large parameters. For instance, Theorem 3.3.3 implies

$$\begin{aligned}&\lim_{N\to\infty} 2\pi^2 R^3 \int_0^1 F(Rx, 2\pi N x, 2\pi N^2 x) x^2 |\sin(2\pi N x)| dx\\ =\ & \textstyle\int_\sigma F(r, \phi, \theta) d\sigma,\end{aligned}$$

which can be considered as an extension of Maréchal–Wilkins' formula (3.1.2) in the three dimensional setting.

Obviously, the line integrals in Theorems 3.3.3 and 3.3.4 are continuous functions of x. Therefore, if a large enough parameter N is chosen so that $R\omega(F; N^{-1}, N^{-1}, 0) + C_F N^{-1}$ or $R\omega(F; N^{-1}, 0, N^{-1}) + C_F N^{-1}$ is small enough, we may use these line integrals to evaluate the triple integral over the sphere domain σ. In addition, if we use Theorem 3.3.1, then we obtain the following DRE with large parameters μ and λ.

$$\begin{aligned} & \int_\sigma F(r,\phi,\theta)d\sigma \\ \approx \; & 2\pi^2 R^3 \int_0^1 F(Rx, 2\pi\lambda x, 2\pi\mu x)x^2|\sin(2\pi\lambda x)|dx; \end{aligned}$$

here we choose the large enough λ first and then μ.

3.4 Basic expansion theorem for integrals with large parameters

In this section, we will discuss the asymptotic expansion of integrals with the form

$$I(f;\lambda) := \int_0^1 f(x, \langle\lambda x\rangle)dx, \tag{3.4.1}$$

where $f(\cdot, y)$ is a periodic function of y with period 1, and λ is a positive real number that may take very large values. Let r be a positive integer. We will show that $I(f;\lambda)$, defined by (3.4.1), may be represented approximately by some easily computed integrals with an error term $O(\lambda^{-r})$ as $\lambda \to \infty$. For large λ, the constant implied by $O(\cdot)$ may be evaluated explicitly.

We will use *Bernoulli polynomials* $B_n(x), n = 0, 1, \ldots$, to give the expansion formulas of the integral (3.4.1). $B_n(x), n = 0, 1, \ldots$, are given by the expansion.

$$\frac{te^{xt}}{e^t - 1} = \sum_{n=0}^{\infty} \frac{t^n}{n!} B_n(x), \quad |t| < 2\pi. \tag{3.4.2}$$

When $x = 0$, the above expression becomes

$$\frac{t}{e^t - 1} = \sum_{n=0}^{\infty} \frac{t^n}{n!} B_n(0). \tag{3.4.3}$$

Here $B_n(0)$ is called the *Bernoulli number*. By dividing $\frac{t}{e^t-1}$ from both sides of equation (3.4.3) and substituting the Taylor expansion $(e^t - 1)/t = \sum_{j=1}^{\infty} t^{j-1}/(j!)$, we have

$$\begin{aligned} 1 &= \frac{e^t - 1}{t} \sum_{k=0}^{\infty} \frac{t^k}{k!} B_j(0) \\ &= \sum_{j=1}^{\infty} \frac{t^{j-1}}{j!} \sum_{k=0}^{\infty} \frac{t^k}{k!} B_k(0) \\ &= \sum_{n=1}^{\infty} t^{n-1} \sum_{k=0}^{n-1} \frac{B_k(0)}{k!(n-k)!}. \end{aligned}$$

Comparing the two sides of the last equation, we obtain the recurrence relation

$$B_0(0) = 1, \quad \sum_{k=0}^{n-1} \frac{1}{k!(n-k)!} B_k(0) = 0 (n \geq 2).$$

After finding Bernoulli numbers, we can use the formula

$$B_n(x) = \sum_{k=0}^{n} \binom{n}{k} B_k(0) x^{n-k} \tag{3.4.4}$$

to find expressions for $B_n(x)$. Formula (3.4.4) is derived from equations (3.4.2) and (3.4.3). In fact, equation (3.4.3) gives

$$\begin{aligned}
\frac{te^{xt}}{e^t-1} &= \frac{t}{e^t-1} e^{xt} \\
&= \sum_{k=0}^{\infty} \frac{t^k}{k!} B_k(0) \cdot \sum_{j=0}^{\infty} \frac{t^j}{j!} x^j \\
&= \sum_{n=0}^{\infty} \frac{t^n}{n!} \cdot \sum_{k=0}^{n} \binom{n}{k} B_k(0) x^{n-k}.
\end{aligned}$$

Comparing the above equation with equation (3.4.2), we immediately have formula (3.4.4). Taking the derivative of equation (3.4.4) p times, we obtain

$$\frac{d^p}{dx^p} B_n(x) = \frac{n!}{(n-p)!} B_{n-p}(x).$$

In particular, for $p = 1$,

$$\frac{d}{dx} B_n(x) = n B_{n-1}(x).$$

When we change n to $n+1$ in the above equation and evaluate the integral, we have

$$\int_a^x B_n(y) dy = \frac{1}{n+1} \left(B_{n+1}(x) - B_{n+1}(a) \right). \tag{3.4.5}$$

Formula (3.4.4) gives expressions of $B_n(x)$: $B_0(x) = 1$, $B_1(x) = x - \frac{1}{2}$, $B_2(x) = x^2 - x + \frac{1}{6}$, $B_3(x) = x^3 - \frac{3}{2}x^2 + \frac{1}{2}x$, ...

Similarly, we substitute the Taylor expansion of $te^x t$ in terms of t and formula (3.4.2) into the relation $\frac{te^{(x+y)t}}{e^t-1} = \frac{te^{yt}}{e^t-1} e^{xt}$ and compare the series in the resulting equation. We thus obtain

$$B_n(x+y) = \sum_{k=0}^{n} \binom{n}{k} B_k(x) y^{n-k}.$$

Repeating the above process for the relation $\frac{te^{(x+1)t}}{e^t-1} = te^{xt} + \frac{te^{xt}}{e^t-1}$, we have

$$B_n(x+1) = B_n(x) + nx^{n-1} \tag{3.4.6}$$

for $n = 0, 1, \ldots$. In addition, applying formula (3.4.2) to both sides of the equation

$$\frac{te^{(1-x)t}}{e^t - 1} = \frac{-te^{-xt}}{e^{-t} - 1}$$

and comparing the series on both sides of the resulting equation, we have

$$B_n(1-x) = (-1)^n B_n(x). \tag{3.4.7}$$

Finally, we apply formula (3.4.6) to equation (3.4.7), then

$$B_n(-x) = (-1)^n \left[B_n(x) + nx^{n-1}\right]. \tag{3.4.8}$$

Denote by $\bar{B}_n(x)$ the *Bernoulli function*, which is a periodic function with period 1. For $x \in [0, 1)$, $\bar{B}_n(x)$ coincides with the nth degree Bernoulli polynomial, $B_n(x)$. Hence, $\bar{B}_n(x) = B_n(\langle x \rangle)$ holds for all real numbers x. Both Bernoulli polynomials and Bernoulli functions will be used in the rest of this section and the next section.

Theorem 3.4.1 (Basic Expansion Theorem) *Assume that the continuous function $f(x, y)$ defined on the domain $[0, 1] \times (-\infty, \infty)$ has the continuous partial derivative $f_x^{(r)}(x, y) \equiv \partial^r f / \partial x^r$. In addition, $f(\cdot, y)$ is a periodic function of y with period 1. Then*

$$\begin{aligned}
&\int_0^1 f(x, \lambda x)dx \\
&= \int_0^1 \int_0^1 f(x, y)dxdy + \sum_{j=1}^{r} \frac{\lambda^{-j}}{j!} \left[\int_{\lambda}^{\lambda+1} f_x^{(j-1)}(1, y) B_j(y - \lambda)dy \right. \\
&\quad \left. - \int_0^1 f_x^{(j-1)}(0, y) B_j(y)dy \right] \\
&\quad - \frac{\lambda^{-r}}{r!} \int_0^1 dx \int_{\lambda x}^{\lambda x+1} f_x^{(r)}(x, y) B_r(y - \lambda x)dx,
\end{aligned} \tag{3.4.9}$$

where $B_r(t)$ is the Bernoulli polynomial of degree r.

Proof. Denote

$$a_r(x) = \frac{\lambda^{-j}}{j!} \int_{\lambda x}^{\lambda x+1} f_x^{(j-1)}(x, y) B_j(y - \lambda x)dy. \tag{3.4.10}$$

We have

$$a_r(1) - a_r(0) = \frac{\lambda^{-j}}{j!}\left[\int_\lambda^{\lambda+1} f_x^{(j-1)}(1, y)B_j(y-\lambda)dy - \int_0^1 f_x^{(j-1)}(0, y)B_j(y)dy\right]. \tag{3.4.11}$$

On the other hand, from the Leibniz differential rule of integrals with parameters, we have

$$a_r'(x) = \frac{\lambda^{-j}}{j!}\int_{\lambda x}^{\lambda x+1} f_x^{(j)}(x, y)B_j(y-\lambda x)dy - \frac{\lambda^{-j+1}}{(j-1)!}\int_{\lambda x}^{\lambda x+1} f_x^{(j-1)}(x, y)B_{j-1}(y-\lambda x)dy + \frac{\lambda^{-j+1}}{j!}\left[f_x^{(j-1)}(x, \lambda x+1)B_j(1) - f_x^{(j-1)}(x, \lambda x)B_j(0)\right]. \tag{3.4.12}$$

Because $f_x^{(j-1)}(\cdot, y)$ is a periodic function of y with period 1 and $B_j(1) = B_j(0)$ for $j \geq 2$ and $B_1(1) - B_1(0) = 1$ (see equation (3.4.6)), the last two terms in the bracket of the above equation can be rewritten as

$$\frac{\lambda^{-j+1}}{j!}\left[f_x^{(j-1)}(x, \lambda x+1)B_j(1) - f_x^{(j-1)}(x, \lambda x)B_j(0)\right] = \begin{cases} 0, & \text{for } 2 \leq j \leq r, \\ f(x, \lambda x), & \text{for } j = 1. \end{cases}$$

Therefore,

$$\begin{aligned}\sum_{j=1}^{r}[a_r(1) - a_r(0)] &= \textstyle\sum_{j=1}^{r}\int_0^1 a_j'(x)dx \\ &= -\int_0^1 dx\int_{\lambda x}^{\lambda x+1} f(x, y)dy + \int_0^1 f(x, \lambda x)dx \\ &\quad + \frac{\lambda^{-r}}{r!}\int_0^1 dx\int_{\lambda x}^{\lambda x+1} f_x^{(r)}(x, y)B_r(y-\lambda x)dy.\end{aligned}$$

Since $f(\cdot, y)$ is a periodic function of y with period 1, the above equation gives

$$\begin{aligned}\int_0^1 f(x, \lambda x)dx &= \int_0^1\int_0^1 f(x, y)dxdy + \sum_{j=1}^{r}[a_r(1) - a_r(0)] \\ &\quad - \frac{\lambda^{-r}}{r!}\int_0^1 dx\int_{\lambda x}^{\lambda x+1} f_x^{(r)}(x, y)B_r(y-\lambda x)dy.\end{aligned}$$

Substituting equation (3.4.11) into the above equation, we thus obtain formula (3.4.9). □

Theorem 3.4.2 *Let $f(x, y)$ be a continuous function defined on the square $V_2 = [0, 1]^2$ that satisfies $\partial^r f/\partial x^r \in C(V_2)$. Then integral $I(f; \lambda)$ defined by (3.4.1) has the expansion*

$$\begin{aligned} I(f;\lambda) &= \int_0^1 \int_0^1 f(x,y)dxdy \\ &+ \sum_{j=1}^r \frac{\lambda^{-j}}{j!} \int_0^1 \left[f_x^{(j-1)}(1,y)\bar{B}_j(y-\lambda) - f_x^{(j-1)}(0,y)B_j(y) \right] dy \\ &- \frac{\lambda^{-r}}{r!} \int_0^1 \int_0^1 f_x^{(r)}(x,y)\bar{B}_r(y-\lambda x)dxdy, \end{aligned} \tag{3.4.13}$$

where $\bar{B}_j(t) = B_j(\langle t \rangle)$ is the Bernoulli function of degree j.

Proof. We separate the proof into two steps. First, if $f(x, 0) = f(x, 1)$, then $f(x, \langle y \rangle)$ satisfies all conditions of Theorem 3.4.1. Therefore, from equation (3.4.9), we have

$$\begin{aligned} & I(f;\lambda) \\ = & \int_0^1 \int_0^1 f(x,y)dxdy + \sum_{j=1}^r \frac{\lambda^{-j}}{j!} \left[\int_\lambda^{\lambda+1} f_x^{(j-1)}(1,\langle y \rangle)\bar{B}_j(y-\lambda)dy \right. \\ & \left. - \int_0^1 f_x^{(j-1)}(0,\langle y \rangle)B_j(y)dy \right] \\ & - \frac{\lambda^{-r}}{r!} \int_0^1 dx \int_{\lambda x}^{\lambda x+1} f_x^{(r)}(x,\langle y \rangle)\bar{B}_r(y-\lambda x)dy. \end{aligned} \tag{3.4.14}$$

In equation (3.4.14), some Bernoulli polynomials in the integrals have been changed to Bernoulli functions because the integral variables are in (0, 1). Therefore, the integrands in the integrals $\int_\lambda^{\lambda+1} \cdots dy$ and $\int_{\lambda x}^{\lambda x+1} \cdots dy$ are periodic functions with period 1. We thus can rewrite these integrals as integrals with the form $\int_0^1 \cdots dy$, and all $\langle y \rangle$ in the integrands can be changed to y. Hence, equation (3.4.14) becomes equation (3.4.13).

Secondly, if $f(x, 0) = f(x, 1)$ is not satisfied, then $f(x, \langle y \rangle)$ is piecewise continuous on $[0, 1] \times (-\infty, \infty)$. The proof of Theorem 3.4.1 still works with the following modification. When we find $a_r'(x)$ (see equation (3.4.12)), for $x \in \left(\frac{i-1}{\lambda}, \frac{i}{\lambda}\right) \cap (0, 1)$ $(i = 1, 2, \dots, i < \lambda + 1)$, we may separate $a_r(x)$ into a sum of two integrals $\int_{\lambda x}^{i} \cdots dy + \int_i^{\lambda x+1} \cdots dy$. Then, by using the Leibniz rule for each integral, we obtain formula (3.4.13). □

Theorem 3.4.3 *Under all conditions of Theorem 3.4.1, when the real parameter* $\lambda \to \infty$,

$$\begin{aligned} I(f;\lambda) &= \int_0^1 \int_0^1 f(x,y)dxdy \\ &\quad + \sum_{j=1}^{r} \frac{\lambda^{-j}}{j!} \int_0^1 \Big[f_x^{(j-1)}(1,y)\bar{B}_j(y-\lambda) \\ &\quad - f_x^{(j-1)}(0,y)B_j(y) \Big] + 0(\lambda^{-r}). \end{aligned} \tag{3.4.15}$$

Proof. From Hsu's *generalization of the Riemann–Lebesque Theorem* (see the proposition 154 in [36], p. 262), we get

$$\begin{aligned} &\lim_{\lambda\to\infty} \left[-\frac{1}{r!} \int_0^1 \int_0^1 f_x^{(r)}(x,y)\bar{B}_r(y-\lambda x)dxdy \right] \\ &= -\tfrac{1}{r!} \int_0^1 \int_0^1 \int_0^1 f_x^{(r)}(x,y)\bar{B}_r(y-u)dudxdy = 0. \end{aligned}$$

The last step is due to formula (3.4.5). Therefore, an application of the above limit to expression (3.4.13) yields formula (3.4.15). □

For continuous functions, $\phi(y)$ and $g(x,y)$, which are defined on $[0,1]$ and $[0,1]\times[0,1]$, respectively, denote

$$\| \phi \| = \left(\int_0^1 (\phi(y))^2 dy \right)^{1/2} \quad \text{and} \quad \| g \| = \left(\int_0^1 \int_0^1 (g(x,y))^2 dxdy \right)^{1/2}.$$

Theorem 3.4.4 *Under conditions of Theorem 3.4.2, denote*

$$\begin{aligned} &S_{r-1}(f;\lambda) \\ &= \int_0^1 \int_0^1 f(x,y)dxdy + \sum_{j=1}^{r-1} \frac{\lambda^{-j}}{j!} \int_0^1 \Big[f_x^{(j-1)}(1,y)\bar{B}_j(y-\lambda) \\ &\quad - f_x^{(j-1)}(0,y)B_j(y) \Big] dy. \end{aligned} \tag{3.4.16}$$

Then for any $\lambda > 0$ *there exists*

$$\begin{aligned} &\left| \int_0^1 f(x,\langle \lambda x \rangle)dx - S_{r-1}(f;\lambda) \right| \\ &\leq \frac{1}{\lambda^r} \left[\left(2(-1)^{r-1} \left(\frac{B_{2r}(0)}{(2r)!} + \frac{B_{2r+1}(\langle -\lambda \rangle)}{(2r+1)!\lambda} \right) \right)^{1/2} \| f_x^{(r)} \| \right. \\ &\quad \left. + \left(2(-1)^{r-1} \frac{B_{2r}(0) - B_{2r}(\langle -\lambda \rangle)}{(2r)!} \right)^{1/2} \| f_x^{(r-1)}(1,\cdot) \| \right], \end{aligned} \tag{3.4.17}$$

where $\| f_x^{(r)} \| = \| f_x^{(r)}(x, y) \|$.

In addition, if $\lambda = N$ *is an integer, then the right-hand side of expression (3.4.17) can be simplified to*

$$N^{-r}\left(2(-1)^{N-1}\frac{B_{2N}}{(2N)!}\right)^{1/2} \| f_x^{(N)} \| .$$

Proof. From equations (3.4.13) and (3.4.14),

$$\begin{aligned} & I(f;\lambda) - S_{r-1}(f;\lambda) \\ = & -\frac{\lambda^{-r}}{r!}\int_0^1\int_0^1 f_x^{(r)}(x,y)\left[\bar{B}_r(y-\lambda x) - B_r(y)\right]dxdy \\ & +\frac{\lambda^{-r}}{r!}\int_0^1 f_x^{(r-1)}(1,y)\left[\bar{B}_r(y-\lambda) - B_r(y)\right]dy. \end{aligned}$$

Applying the Cauchy–Schwartz inequality to two terms on the right-hand side of the above equation and evaluating it as follows, we can obtain inequality (3.4.17).

$$\begin{aligned} & \frac{1}{(r!)^2}\int_0^1\int_0^1 \left(\bar{B}_r(y-\lambda x) - B_r(y)\right)^2 dxdy \\ = & \frac{1}{(r!)^2}\int_0^1 dx\int_0^1 \left(\bar{B}_r(y-\lambda x)^2 - 2\bar{B}_r(y-\lambda x)B_r(y) + B_r(y)^2\right)dy \\ = & \frac{2}{(r!)^2}\int_0^1 dx\int_0^1 \left(B_r(y) - \bar{B}_r(y-\lambda x)\right)B_r(y)dy \\ = & 2\int_0^1\left\{\sum_{j=0}^{r-1}(-1)^j\left[\frac{B_{r+j+1}(y) - \bar{B}_{r+j+1}(y-\lambda x)}{(r+j-1)!}\frac{B_{r-j}(y)}{(r-j)!}\right]_{y=0}^{y=1}\right. \\ & \left. +(-1)^r\int_0^1 \frac{B_{2r}(y) - \bar{B}_{2r}(y-\lambda x)}{(2r)!}dy\right\}dx \\ = & 2(-1)^{r-1}\int_0^1 \frac{B_{2r}(0) - \bar{B}_{2r}(-\lambda x)}{(2r)!}dx \\ = & 2(-1)^{r-1}\left[\frac{B_{2r}(0)}{(2r)!} + \frac{\bar{B}_{2r+1}(-\lambda)}{(2r+1)!\lambda}\right]. \end{aligned}$$

$$\begin{aligned} & \frac{1}{(r!)^2}\int_0^1 \left(\bar{B}_r(y-\lambda) - B_r(y)\right)^2 dy \\ = & \frac{2}{(r!)^2}\int_0^1 \left(B_r(y) - \bar{B}_r(y-\lambda)\right)B_r(y)dy \\ = & 2(-1)^{r-1}\frac{B_{2r}(0) - \bar{B}_{2r}(-\lambda)}{(2r)!}. \end{aligned}$$

□

Theorem 3.4.5 *Let $\phi(x) \in C^r([0,1])$ and $\psi(x)$ be a continuous periodic function with period 1. Then, for*

$$I = \int_0^1 \phi(x)\psi(\lambda x)dx$$

and

$$\begin{aligned} S_r &= \int_0^1 \phi(x)dx \int_0^1 \psi(x)dx + \sum_{j=1}^r \frac{\lambda^{-j}}{j!}\Big[\phi^{(j-1)}(1) \\ &\quad \times \int_\lambda^{\lambda+1} \psi(x)B_j(x-\lambda)dx - \phi^{(j-1)}(0)\int_0^1 \psi(x)B_j(x)dx\Big], \end{aligned}$$

we have

$$I = S_r + O\left(\lambda^{-r}\right)$$

as $\lambda \to \infty$, where

$$O\left(\lambda^{-r}\right) = -\frac{\lambda^{-r}}{r!}\int_0^1 \phi^{(r)}(x)\int_{\lambda x}^{\lambda x+1} \psi(t)B_r(t-\lambda x)dtdx.$$

In addition,

$$I = S_{r-1} + O\left(\lambda^{-r}\right),$$

where

$$\begin{aligned} &|O\left(\lambda^{-r}\right)| \\ &\le \lambda^{-r}\left[\left(2(-1)^{r-1}\left(\frac{B_{2r}}{(2r)!} + \frac{B_{2r+1}(\langle -\lambda\rangle)}{(2r+1)!\lambda}\right)\right)^{1/2} \| \phi^{(r)} \| \right. \\ &\left. + \left(2(-1)^{r-1}\frac{B_{2r}(0) - B_{2r}(\langle -\lambda\rangle)}{(2r)!}|\phi^{(r-1)}(1)|\right)^{1/2} \| \psi \| \right] \quad (3.4.18) \end{aligned}$$

Proof. By replacing $f(x,y)$ with $\phi(x)\psi(y)$ in Theorems 3.4.1, 3.4.3, and 3.4.4, we obtain the results in this theorem. □

As an example of the application of Theorem 3.4.5, let us consider the integral (see [42])

$$I = \int_0^1 \cos(2\pi x)\sin(25\pi x)dx.$$

We take $\phi(x) = \cos(2\pi x)$, $\psi(x) = \sin(25\pi x)$, $\lambda = 12.5$, and $r = 2m+1$ $(m \in N)$ in Theorem 3.4.5. Thus,

$$\begin{aligned} &S_{2m} \\ &= \sum_{j=1}^m \frac{2(-1)^j(2\pi)^{2j-2}}{(2j-1)!}\left(\frac{2}{25}\right)^{2j-1}\int_0^1 \sin(2\pi x)B_{2j-1}(x)dx \\ &= \frac{1}{\pi}\sum_{j=1}^m \left(\frac{2}{25}\right)^{2j-1}, \end{aligned}$$

for $m = 1, 2, 3, \ldots$. Because

$$\| \phi^{(r)} \| = 2^{r-\frac{1}{2}}\pi^r, \qquad \| \psi \| = \frac{1}{\sqrt{2}},$$

and

$$\frac{|B_{2r}(0)|}{(2r)!} = \frac{1}{2^{2r-1}\pi^{2r}} \sum_{j=1}^{\infty} \frac{1}{j^{2r}} \approx \frac{1}{2^{2r-1}\pi^{2r}},$$

we have

$$|I - S_{2m}| \leq \left(\frac{2}{25}\right)^{2m+1} \left(1 + \frac{1}{\pi}\right).$$

Therefore,

$$|I - S_2| < 0.00067, \qquad |I - S_4| < 0.0000044,$$

etc. In fact,

$$S_2 = \frac{2}{25\pi} = 0.0254648, \quad S_4 = \left(1 + \left(\frac{2}{25}\right)^2\right) \frac{2}{25\pi} = 0.0256278,$$

and

$$I = \left(1 - \left(\frac{2}{25}\right)^2\right)^{-1} \frac{2}{25\pi} = 0.0256288.$$

Hence,

$$I - S_2 = 0.00016, \qquad I - S_4 = 0.0000010.$$

If λ is an integer N, from Theorem 3.4.1 we have the following result.

Theorem 3.4.6 *Let $f(x, y)$ be a continuous function defined on the domain $D = \{(x, y) : 0 \leq x \leq 1, -\infty < y < \infty\}$ with $f_x^{(r)}(x, y) \in C(D)$ and $f(\cdot, y)$ being a periodic function in y with period 1. Then there exists the expansion*

$$\begin{aligned} \int_0^1 f(x, Nx)dx &= \int_0^1 \int_0^1 f(x, y)dxdy \\ &\quad + \sum_{j=1}^{r} \frac{1}{j!} \left(\frac{1}{N}\right)^j \int_0^1 \left[\frac{\partial^{j-1} f}{\partial x^{j-1}}\right]_{x=0}^{x=1} B_j(y)dy + \rho_r, \end{aligned}$$

where the remainder ρ_r can be written as

$$\rho_r = -\frac{1}{r!} \left(\frac{1}{N}\right)^r \int_0^1 dx \int_{Nx}^{Nx+1} B_r(y - Nx) \left(\frac{\partial^r f}{\partial x^r}\right) dy$$

and has the estimate

$$|\rho_r| \leq \left(\frac{1}{N}\right)^r \left(2(-1)^{r-1}\frac{B_{2r}(0)}{(2r)!}\right)^{1/2} \left\|\frac{\partial^r f}{\partial x^r}\right\|_C .$$

We will apply Theorem 3.4.6 to derive numerical quadrature formulas for a type of double integrals in Section 4.1.

3.5 Asymptotic expansion formulas for oscillatory integrals with singular factors

We now establish asymptotic expansion formulas for oscillatory integrals of the form

$$I(f;\lambda,\mu) = \int_0^1 x^{-\mu} f(x, \langle\lambda x\rangle)dx, \tag{3.5.1}$$

where $0 < \mu < 1$ and λ is a large real parameter.

We need the following two lemmas.

Lemma 3.5.1 *Let* $0 \leq y \leq 1$, $\lambda > 0$, *and* $\phi(x) \in C^r([0,1])$, $r \geq 1$. *Then*

$$\begin{aligned} &\frac{1}{\lambda}\sum_{k=0}^{[\lambda-y]} \phi\left(\frac{k+y}{\lambda}\right) \\ &= \int_0^1 \phi(x)dx + \sum_{j=1}^{r} \frac{\lambda^{-j}}{j!}\left\{\bar{B}_j(y-\lambda)\phi^{(j-1)}(1)\right. \\ &\quad \left. -B_j(y)\phi^{(j-1)}(0)\right\} - \frac{\lambda^{-r}}{r!}\int_0^1 \bar{B}_r(y-\lambda x)\phi^{(r)}(x)dx, \end{aligned} \tag{3.5.2}$$

where the upper limit of the first sum, $[\lambda - y]$, *is the largest integer that is less than or equal to* $\lambda - y$. $\bar{B}_j(y)$, *as usual, is the Bernoulli function with period* 1 *and* $\bar{B}_j(y) = B_j(\langle y\rangle)$.

Proof. Applying integration by parts to the last integral in the above equation, we know formula (3.5.2) to be correct. □

Similarly, we have

$$\begin{aligned} &\frac{1}{\lambda}\sum_{k=1}^{[\lambda-y]} \phi\left(\frac{k+y}{\lambda}\right) \\ &= \int_{\frac{1}{\lambda}}^1 \phi(x)dx + \sum_{j=1}^{r} \frac{\lambda^{-j}}{j!}\left\{\bar{B}_j(y-\lambda)\phi^{(j-1)}(1)\right. \\ &\quad \left. -B_j(y)\phi^{(j-1)}(0)\right\} - \frac{\lambda^{-r}}{r!}\int_{1/\lambda}^1 \bar{B}_r(y-\lambda x)\phi^{(r)}(x)dx. \end{aligned} \tag{3.5.3}$$

Lemma 3.5.2 *Assume r to be a positive integer, $s > -r (s \neq 1)$, and $0 < y \leq 1$. Then*

$$\begin{aligned}
&\sum_{k=0}^{[\lambda-y]} \frac{1}{(k+y)^s} \\
=\ & \zeta(s, y) + \frac{\lambda^{1-s}}{1-s} + \lambda^{1-s} \sum_{j=1}^{r} \binom{-s}{j-1} \frac{\bar{B}_j(y-\lambda)}{j\lambda^j} \\
& + \binom{-s}{r} \int_{\lambda}^{\infty} \bar{B}_r(y-x) x^{-r-s} dx, \qquad (3.5.4)
\end{aligned}$$

where $\zeta(s, y)$ ($s > -r, s \neq 1, 0 < y \leq 1$) is the generalized Riemann–Zeta function.

Proof. Denote $\phi(x) = x^{-s}$. For $0 < y \leq 1$, from formula (3.5.3),

$$\begin{aligned}
& \tfrac{1}{\lambda} \textstyle\sum_{k=0}^{[\lambda-y]} \left(\frac{k+y}{\lambda}\right)^{-s} \\
=\ & \tfrac{1}{\lambda} \left(\tfrac{y}{\lambda}\right)^{-s} + \tfrac{1}{\lambda} \textstyle\sum_{k=1}^{[\lambda-y]} \left(\frac{k+y}{\lambda}\right)^{-s} \\
=\ & \lambda^{s-1} y^{-s} \int_{\frac{1}{\lambda}}^{1} x^{-s} dx + \textstyle\sum_{j=1}^{r} \binom{-s}{j-1} \frac{\bar{B}_j(y-\lambda) - B_j(y)\lambda^{j-1+s}}{j\lambda^j} \\
& - \tfrac{1}{\lambda^r} \binom{-s}{r} \int_{\frac{1}{\lambda}}^{1} \bar{B}_r(y - \lambda x) x^{-r-s} dx.
\end{aligned}$$

Therefore,

$$\begin{aligned}
&\sum_{k=0}^{[\lambda-y]} \frac{1}{(k+y)^s} \\
=\ & y^{-s} - \frac{1}{1-s} - \sum_{j=1}^{r} \binom{-s}{j-1} \frac{B_j(y)}{j} \\
& - \binom{-s}{r} \int_{1}^{\infty} \bar{B}_r(y-x) x^{-r-s} dx + \frac{\lambda^{1-s}}{1-s} + \lambda^{1-s} \sum_{j=1}^{r} \binom{-s}{j-1} \\
& \times \frac{\bar{B}_j(y-\lambda)}{j\lambda^j} + \binom{-s}{r} \int_{\lambda}^{\infty} \bar{B}_r(y-x) x^{-r-s} dx. \qquad (3.5.5)
\end{aligned}$$

If $s > 1$ and $\lambda \to \infty$, then

$$\begin{aligned}
& \zeta(s, y) \\
=\ & y^{-s} - \frac{1}{1-s} - \sum_{j=1}^{r} \binom{-s}{j-1} \frac{B_j(y)}{j} \\
& - \binom{-s}{r} \int_{1}^{\infty} \bar{B}_r(y-x) x^{-r-s} dx. \qquad (3.5.6)
\end{aligned}$$

Note that the infinite integral in expression (3.5.6) converges if $0 < y \leq 1$ and $s > -r(s \neq 1)$. Hence, $\zeta(s, y)$ is an analytic extension of the Riemann–Zeta function on $0 < y \leq 1$, $s > -r(s \neq 1)$. □

Now we give a *generalization of the Euler–Maclaurin summation formula*. Let $0 < s < 1$, $0 < y \leq 1$, and $\psi(x) = x^s\phi(x) \in C^{r+1}([0, 1])$. Then

$$\Phi(x) = \phi(x) - \sum_{j=0}^{r} \frac{\psi^{(j)}(0)}{j!} x^{j-s} \in C_0^r([0, 1]).$$

Therefore, from equation (3.5.2),

$$\begin{aligned}
&\frac{1}{\lambda}\sum_{k=0}^{[\lambda-y]} \Phi\left(\frac{k+y}{\lambda}\right) \\
= \ & \frac{1}{\lambda}\sum_{k=0}^{[\lambda-y]} \left\{\phi\left(\frac{k+y}{\lambda}\right) - \sum_{i=0}^{r} \frac{\psi^{(i)}(0)}{i!}\left(\frac{k+y}{\lambda}\right)^{i-s}\right\} \\
= \ & \int_0^1 \phi(x)dx - \sum_{i=0}^{r} \frac{1}{i+1-s}\frac{\phi^{(i)}(0)}{i!} \\
& + \sum_{j=1}^{r} \frac{\bar{B}_j(y-\lambda)}{j\lambda^j}\left\{\frac{\phi^{(j-1)}(1)}{(j-1)!} - \sum_{i=0}^{r} \frac{\psi^{(i)}(0)}{i!}\binom{i-s}{j-1}\right\} \\
& - \frac{1}{r!\lambda^r}\int_0^1 \bar{B}_r(y-\lambda x)\Phi^{(r)}(x)dx;
\end{aligned}$$

that is,

$$\begin{aligned}
&\frac{1}{\lambda}\sum_{k=0}^{[\lambda-y]} \phi\left(\frac{k+y}{\lambda}\right) \\
= \ & \int_0^1 \phi(x)dx + \sum_{j=1}^{r} \frac{\bar{B}_j(y-\lambda)}{j!\lambda^j}\phi^{(j-1)}(1) + \sum_{i=0}^{r} \frac{\psi^{(i)}(0)}{i!}\left(\frac{1}{\lambda}\right)^{i+1-s} \\
\times \ & \left\{\sum_{k=0}^{[\lambda-y]} \frac{1}{(k+y)^{s-i}} - \frac{\lambda^{i+1-s}}{i+1-s} - \lambda^{i+1-s}\sum_{j=1}^{r} \frac{\bar{B}_j(y-\lambda)}{j\lambda^j}\binom{i-s}{j-1}\right\} \\
- \ & \frac{1}{r!\lambda^r}\int_0^1 \bar{B}_r(y-\lambda x)\Phi^{(r)}(x)dx. \qquad (3.5.7)
\end{aligned}$$

An application of (3.5.4) to the expression inside the curly brackets of (3.5.7) yields the following *generalized Euler–Maclaurin formula*.

$$\begin{aligned}
&\frac{1}{\lambda}\sum_{k=0}^{[\lambda-y]}\phi\left(\frac{k+y}{\lambda}\right) \\
&= \int_0^1 \phi(x)dx + \sum_{j=1}^{r}\frac{\bar{B}_j(y-\lambda)}{j!\lambda^j}\phi^{(j-1)}(1) \\
&\quad + \sum_{i=0}^{r}\frac{\psi^{(i)}(0)}{i!}\left(\frac{1}{\lambda}\right)^{i+1-s}\zeta(s-i,y) + R_r,
\end{aligned} \tag{3.5.8}$$

where

$$\begin{aligned}
R_r &= \sum_{i=0}^{r}\frac{\psi^{(i)}(0)}{i!}\left(\frac{1}{\lambda}\right)^{i+1-s}\binom{i-s}{r}\int_\lambda^\infty \bar{B}_r(y-x)x^{i-s-r}dx \\
&\quad -\frac{1}{r!\lambda^r}\int_0^1 \bar{B}_r(y-\lambda x)\Phi^{(r)}(x)dx \\
&= \frac{1}{\lambda^r}\int_1^\infty \bar{B}_r(y-\lambda x)\sum_{i=0}^{r}\frac{\psi^{(i)}(0)}{i!}\binom{i-s}{r}x^{i-s-r}dx \\
&\quad -\frac{1}{r!\lambda^r}\int_0^1 \bar{B}_r(y-\lambda x)\Phi^{(r)}(x)dx \\
&= \frac{1}{r!\lambda^r}\int_1^\infty \bar{B}_r(y-\lambda x)\left[\sum_{i=0}^{r}\frac{\psi^{(i)}(0)}{i!}x^{i-s}\right]^{(r)} dx \\
&\quad -\frac{1}{r!\lambda^r}\int_0^1 \bar{B}_r(y-\lambda x)\Phi^{(r)}(x)dx,
\end{aligned}$$

or equivalently,

$$\begin{aligned}
R_r &= \frac{1}{r!\lambda^r}\int_1^\infty \bar{B}_r(y-\lambda x)[\phi(x)-\Phi(x)]^{(r)}dx \\
&\quad -\frac{1}{r!\lambda^r}\int_0^1 \bar{B}_r(y-\lambda x)\Phi^{(r)}(x)dx.
\end{aligned}$$

Here

$$\Phi(x) = \frac{x^{-s}}{r!}\int_0^1 \psi^{(r+1)}(t)(x-t)^r dt.$$

Now we will use the generalized Euler–Maclaurin formula (3.5.8) to prove the following *expansion formula for oscillatory integrals with singular factors* x^{-s} *(*$0 < s < 1$*).*

Theorem 3.5.3 *Let $F(x, y)$ be a continuous function defined on the square domain $V_1 = [0, 1]^2$, and let $F_x^{(r)}(x, y) \in C([0, 1]^2)$. Denote*

$$\begin{aligned} f(x, y) &= x^{-s} F(x, y), \\ f_r(x, y) &= \sum_{j=0}^{r} \frac{1}{j!} F_x^{(j)}(x, y) x^{j-s}, \end{aligned}$$

and

$$g_r(x, y) = f(x, y) - f_r(x, y),$$

where $0 < s < 1$. Then for any positive real number λ,

$$\begin{aligned} I(f; \lambda) &= \int_0^1 x^{-s} F(x, \langle \lambda x \rangle) dx \\ &= \int_0^1 \int_0^1 f(x, y) dx dy + \sum_{j=1}^{r} \frac{\lambda^{-j}}{j!} \int_0^1 \bar{B}_j(y - \lambda) f_x^{(j-1)}(1, y) dy \\ &\quad + \sum_{j=0}^{r} \frac{\lambda^{-j-1+s}}{j!} \int_0^1 \zeta(s - j, y) F_x^{(j)}(0, y) dy \\ &\quad + \frac{\lambda^{-r}}{r!} \int_0^1 dy \int_1^{\infty} \bar{B}_r(y - \lambda x) f_x^{(r)}(x, y) dx \\ &\quad - \frac{\lambda^{-r}}{r!} \int_0^1 dy \int_0^1 \bar{B}_r(y - \lambda x) g_x^{(r)}(x, y) dx. \end{aligned} \tag{3.5.9}$$

Proof. Because

$$f(x, y) = x^{-s} F(x, y) = \sum_{i=0}^{r} \frac{F_x^{(i)}(0, y)}{i!} x^{i-s} + g_r(x, y),$$

there is

$$\begin{aligned} \int_0^1 x^{-s} F(x, \langle \lambda x \rangle) dx &= \int_0^1 f(x, \langle \lambda x \rangle) dx \\ &= \frac{1}{\lambda} \int_0^{\lambda} f\left(\frac{y}{\lambda}, \langle y \rangle\right) dy \\ &= \frac{1}{\lambda} \int_0^1 \sum_{k=0}^{[\lambda - y]} f\left(\frac{k + y}{\lambda}, y\right) dy. \end{aligned} \tag{3.5.10}$$

On the other hand, from the generalized Euler–Maclaurin formula (3.5.8),

$$
\begin{aligned}
&\frac{1}{\lambda}\sum_{k=0}^{[\lambda-y]} f\left(\frac{k+y}{\lambda}, y\right) \\
= &\int_0^1 f(x,y)dx + \sum_{j=1}^{r} \frac{\bar{B}_j(y-\lambda)}{j!\lambda^j} f_x^{(j-1)}(1,y) \\
&+\sum_{j=0}^{r} \frac{\zeta(s-j,y)}{j!\lambda^{j+1-s}} F_x^{(j)}(0,y) + \frac{\lambda^{-r}}{r!}\int_1^{\infty} \bar{B}_r(y-\lambda x) \\
&\times \left(\frac{\partial}{\partial x}\right)^r [f(x,y)-g_r(x,y)]dx - \frac{\lambda^{-r}}{r!}\int_0^1 \bar{B}_r(y-\lambda x)(g_r)_x^{(r)}(x,y)dx.
\end{aligned}
$$

Substituting the above equation into equation (3.5.10), we immediately obtain expansion (3.5.9). □

Theorem 3.5.3 can be used to estimate oscillatory integrals with singular factors. Here the singular point is at $x = 0$. We can find similar expansions for oscillatory integrals with singularities at other points.

For more detailed expositions on the topics discussed in Sections 3.4 and 3.5, please see [62], [63], [38], [71], and [70].

Chapter 4

Numerical Quadrature Formulas Associated with the Integration of Rapidly Oscillating Functions

In Sections 1 and 2, we will discuss numerical quadrature for a type of double integrals and for oscillatory integrals, respectively. The compound degree of precision of the approximate computation of oscillatory integrals will be looked at in Section 3. Section 4 will give a fast numerical computation technique shown in Bradie, Coifman, and Grossmann [2] for treating model oscillatory boundary integral operators with or without singularities in $\mathbb{R}^2$. Finally, we will apply Burrows' DRE ([6]) to numerical integration in Section 5. Burrows' DRE is an exact DRE, which can reduce multidimensional Lebesgue integrals to one-dimensional integrals of measure functions. This technique is particularly useful for integrands that are highly oscillatory in character or are singular. The corresponding error analysis for finding measure functions will be given in Section 6.

4.1 Numerical quadrature formulas of double integrals

We will use expansion in Theorem 3.4.6 to derive numerical quadrature formulas for a type of double integrals.

Let E be the domain $\{(x, y) : \psi(x) \le y \le \phi(x), 0 \le x \le 1\}$, where $\phi(x)$ and $\psi(x)$ are continuous or differentiable functions. Denote

$$\theta(x) = \phi(x) - \psi(x) \quad (0 \le x \le 1).$$

In addition, for $x > 0$ we still use notation $\langle x \rangle = x - [x]$, where $[x]$ is the integer part of x.

From Section 3.4 and using a suitable linear transform, it is easy to know that the double integral of $f(x, y)$ over the region E, $\iint_E f(x, y)dS$, can be approximated by a single integral with the form

$$\int_0^1 f(x, \psi(x) + \theta(x)\langle Nx \rangle)dx.$$

Here N is a large integer and dS is the area measure. We are in a dilemma: If N is large, then the integrand is strongly oscillatory, and it is discontinuous at points $x = \frac{1}{N}, \frac{2}{N}, \ldots$ as well; if it is small, then the single integral cannot give a good approximation to the double integral. Hence, two problems arise. First, how to change the integrand so that it can be a continuous function. Secondly, how to find additional terms to the single integral such that they can approximate the double integral with a small N. We will show that the first problem can be solved readily and easily. Theorem 3.4.6 can be used to solve the second problem.

It is easy to see

$$\begin{aligned}\iint_E f(x, y)dS &= \int_0^1 dx \int_0^{\theta(x)} f(x, \psi(x) + t)dt \\ &= \int_0^1 dx \int_0^1 f(x, \psi(x) + \theta(x)y)\theta(x)dy.\end{aligned}$$

In order to extend the integrand in the last integral of the above equation to the region $0 \le x \le 1, -\infty < y < \infty$, we may change it to $f(x, \psi(x)+\theta(x)\langle y \rangle)\theta(x)$. But the integrand is discontinuous at points $y = 0, \pm 1, \pm 2, \ldots$. However, we note that

$$\iint_E f(x, y)dS = \int_0^1 dx \int_0^1 f(x, \phi(x) - \theta(x)y)\theta(x)dy$$

and the function

$$\begin{aligned}\Phi(x, y) &= \frac{1}{2}\theta(x)\{f(x, \psi(x) + \theta(x)\langle y \rangle) \\ &\quad + f(x, \phi(x) - \theta(x)\langle y \rangle)\}\end{aligned} \tag{4.1.1}$$

is continuous at points $y = 0, \pm 1, \pm 2, \ldots$. In fact, $\Phi(x, y) \equiv \Phi(x, \langle y \rangle)$ and

$$\lim_{y \to 0+} \Phi(x, y) = \Phi(x, 0) = \Phi(x, 1) = \lim_{y \to 1^-} \Phi(x, y).$$

Thus, we have

$$\iint_E f(x, y)dS = \int_0^1 \int_0^1 \Phi(x, y)dxdy. \tag{4.1.2}$$

Here $\Phi(x, y)$ is defined by equation (4.1.1), which is a continuous function and $\Phi(\cdot, y)$ is a periodic function with period 1. In addition, if $f_x^{(r)}(x, y)$ and $f_y^{(r)}(x, y)$ are continuous and $\phi^{(r)}(x)$, $\psi^{(r)}(x) \in C([0, 1])$, then $\Phi_x^{(r)}(x, y)$ is continuous. Therefore, from Theorem 3.4.6 we have the following result.

Theorem 4.1.1 *Let $f(x, y)$ be a function defined on $E = \{(x, y) : \psi(x) \leq y \leq \phi(x), 0 \leq x \leq 1\}$ with continuous $f_x^{(r)}(x, y)$ and $f_y^{(r)}(x, y)$ and $\phi^{(r)}(x)$, $\psi^{(r)}(x) \in C([0, 1])$. Then we have*

$$\begin{aligned} \iint_E f(x, y)dS &= \int_0^1 \Phi(x, Nx)dx - \sum_{j=1}^{r-1} \frac{1}{j!}\left(\frac{1}{N}\right)^j \\ &\quad \times \int_0^1 \left[\frac{\partial^{j-1}\Phi}{\partial x^{j-1}}\right]_{x=0}^{x=1} B_j(y)dy + \delta_N, \end{aligned} \tag{4.1.3}$$

where $\Phi(x, y)$ is defined by equation (4.1.1) and the remainder δ_N has estimate

$$|\delta_N| \leq \frac{2}{r!}\left(\frac{1}{N}\right)^r \parallel B_r(x) \parallel \parallel \Phi_x^{(r)}(x, y) \parallel . \tag{4.1.4}$$

When we apply formula (4.1.3) to evaluate the double integral, we should choose a suitable integer r such that not too large an N will be taken and, hence, the single integral $\int_0^1 \Phi(x, Nx)dx$ can be evaluated easily.

Formula (4.1.3) is suitable for all types of region E provided that the boundary curves $y = \psi(x)$ and $y = \phi(x)$ are smooth enough. In particular, if $\psi(x) = 0$ and $\phi(x) = 1$, then expansion (4.1.3) is reduced to formula (4.1.3) in Theorem 3.4.6.

We now consider an important type of region, for which the conditions $\psi(0) = \phi(0)$ and $\psi(1) = \phi(1)$ are satisfied. Therefore, for $\theta(x) = \phi(x) - \psi(x)$, there are $\theta(0) = \theta(1) = 0$. In addition, $\psi(x)$ and $\psi(x)$ are differentiable with finite values of $\phi'(0)$, $\phi'(1)$, $\psi'(0)$, and $\psi'(1)$. (Hence, at points $(0, \phi(0))$ and $(1, \phi(1))$, the boundary curve has no vertical tangent lines.) It is obvious that this type of region does not contain those with smooth boundary curves. However, many complicated regions can be split into several regions of this type as well as possibly some simplexes.

In the following, we will assume that $f(x, y)$, $\phi(x)$, and $\psi(x)$ have continuous derivatives (or partial derivatives) of order 4. Hence, $\Phi(x, y)$ has 4th partial derivatives on E. In this case, under the assumption that E is the region bounded by two cross curves (i.e., $\psi(0) = \phi(0)$ and $\psi(1) = \phi(1)$), we have

$$\int_0^1 \left[\Phi_x^{(j-1)}(x, y)\right]_{x=0}^{x=1} B_j(y)dy = 0, \quad j = 1, 2, 3. \tag{4.1.5}$$

Therefore, from Theorem 4.1.1 and estimation (4.1.4), we can prove the following theorem.

Theorem 4.1.2 *Assume that $f(x, y)$ has continuous 4th order partial de-rivatives on $E = \{(x, y) : \psi(x) \le y \le \phi(x), 0 \le x \le 1\}$; $\phi(x)$ and $\psi(x)$ have 4th order continuous derivatives on $[0, 1]$; and $\psi(0) = \phi(0)$, $\psi(1) = \phi(1)$. Then there exists*

$$\iint_E f(x, y)dS = \int_0^1 \Phi(x, Nx)dx + \delta_N, \tag{4.1.6}$$

where the remainder δ_N has estimate

$$|\delta_N| \le \frac{1}{360}\left(\frac{1}{N}\right)^4 \parallel \Phi_x^{(4)}(x, y) \parallel . \tag{4.1.7}$$

Proof. It is sufficient to verify the three equations shown in (4.1.5) and inequality (4.1.7). From equation (3.4.5), we have $\int_0^1 B_j(y)dy = (B_{j+1}(1)-B_{j+1}(0))/(j+1) = 0$ for all $j \ge 1$. Noting $\theta(0) = \theta(1) = 0$, $\phi(0) = \psi(0)$, and $\phi(1) = \psi(1)$, we have $\Phi(1, y) - \Phi(0, y) \equiv 0$, $\Phi'_x(1, y) - \Phi'_x(0, y) = \theta'(1)f(1, \phi(1)) - \theta'(0)f(0, \phi(0))$, and $\Phi''_{xx}(1, y) - \Phi''_{xx}(0, y) =$constant. From the last three equations and $\int_0^1 B_j(y)dy = 0$, we immediately know that equation (4.1.5) holds for $j = 1, 2,$ and 3. Finally, we evaluate the value of $\parallel B_4(y) \parallel$:

$$\begin{aligned}
\parallel B_4(y) \parallel &= \max_{0\le y\le 1}\left|y^4 - 2y^3 + y^2 - \frac{1}{30}\right| \\
&= \max_{0\le y\le 1}\left|y^2(1-y)^2 - \frac{1}{30}\right| \\
&= \frac{1}{30}.
\end{aligned}$$

From the above value of $\parallel B_4(y) \parallel$ and equations (4.1.3), (4.1.4), and (4.1.5), we obtain formula (4.1.6) and remainder estimate (4.1.7). □

Example 4.1.1. Take $N = 10$, from (4.1.6) and (4.1.7),

$$\iint_E f(x, y)dS = \frac{1}{10}\int_0^1 \left\{\sum_{j=0}^{9} \Phi\left(\frac{x+j}{10}, x\right)\right\} + \delta_{10}, \tag{4.1.8}$$

where $\Phi(x, y)$ is defined by (4.1.1) and δ_{10} has estimate

$$|\delta_{10}| \le \frac{1}{3600000} \parallel \Phi_x^{(4)}(x, y) \parallel . \tag{4.1.9}$$

As for the definite integrals in equation (4.1.8), they can be evaluated by using suitable numerical quadrature formulas, e.g., Gaussian formulas.

In order to modify the results in Theorem 4.1.2, let us evaluate the integral shown in (4.1.5) for $j = 4$. First, we have

$$\int_0^1 yB_4(y)dy = \int_0^1 B_4(y)dy = 0 \tag{4.1.10}$$

and

$$\int_0^1 y^2 B_4(y)dy = -\frac{1}{630}. \tag{4.1.11}$$

Secondly, we compute the third partial derivative values of $\Phi(x, y)$ in terms of x at points $x = 0$ and 1 as

$$\Phi_x^{(3)}(1, y) = 3(\theta'(1))^3 f''_{yy}(1, \phi(1))y^2 + Ay + constant \tag{4.1.12}$$

and

$$\Phi_x^{(3)}(0, y) = 3(\theta'(0))^3 f''_{yy}(0, \phi(0))y^2 + A'y + constant \tag{4.1.13}$$

where A and A' are two constants. Applying Theorem 4.1.1 and using equations (4.1.10)–(4.1.13), we obtain the following theorem.

Theorem 4.1.3 *Assume $f(x, y)$, $\phi(x)$, and $\psi(x)$ have continuous 5th order derivatives (or partial derivatives). Then*

$$\begin{aligned}\iint_E f(x, y)dS &= \int_0^1 \Phi(x, Nx)dx + \frac{1}{5040}\left\{f''_{yy}(1, \phi(1))(\theta'(1))^3\right.\\ &\left.-f''_{yy}(0, \phi(0))(\theta'(0))^3\right\}\left(\frac{1}{N}\right)^4 + \epsilon_N,\end{aligned} \tag{4.1.14}$$

where the remainder ϵ_N has estimate

$$|\epsilon_N| < \frac{1}{1500}\left(\frac{1}{N}\right)^5 \parallel \Phi_x^{(5)} \parallel. \tag{4.1.15}$$

Proof. Equation (4.1.14) is obvious; let us prove estimation (4.1.15), which comes from $\parallel B_5(y) \parallel, 0 \le y \le 1$. Because $B_5(0) = B_5(1) = 0$, we have

$$\parallel B_5(y) \parallel = \max_{0\le y\le 1}\left|y^5 - \frac{5}{2}y^4 + \frac{5}{3}y^3 - \frac{1}{6}y\right|.$$

It is easy to find that the maximum of $|B_5(y)|$ happens at points

$$y_1 = \frac{1}{2} + \sqrt{1 - \frac{2}{15}\sqrt{30}}, \quad y_2 = \frac{1}{2} - \sqrt{1 - \frac{2}{15}\sqrt{30}}.$$

Since $\max\{|B_5(y_1)|, |B_5(y_2)|\} < 1/25$, from (4.1.4) we obtain inequality (4.1.15). □

Applying Theorems 4.1.2 and 4.1.3, we can construct some numerical quadrature formulas. In the following, we will give two examples.

Example 4.1.2. In Theorem 4.1.3, we take $N = 3$ and apply Simpson's composite rule of 11 nodes to the single integral, obtaining the approximation formula

$$\begin{aligned}
&\iint_E f(x, y)dS \\
\approx\; & \tfrac{2}{15}\left\{\Phi\left(\tfrac{1}{10}, \tfrac{3}{10}\right) + \Phi\left(\tfrac{3}{10}, \tfrac{9}{10}\right) + \Phi\left(\tfrac{1}{2}, \tfrac{1}{2}\right)\right. \\
& \left.+\Phi\left(\tfrac{7}{10}, \tfrac{1}{10}\right) + \Phi\left(\tfrac{9}{10}, \tfrac{7}{10}\right)\right\} + \tfrac{1}{15}\left\{\Phi\left(\tfrac{1}{5}, \tfrac{3}{5}\right)\right. \\
& \left.+\Phi\left(\tfrac{2}{5}, \tfrac{1}{5}\right) + \Phi\left(\tfrac{3}{5}, \tfrac{4}{5}\right) + \Phi\left(\tfrac{4}{5}, \tfrac{2}{5}\right)\right\} \\
& +\tfrac{1}{408240}\left\{f''_{yy}(1, \phi(1))(\theta'(1))^3 - f''_{yy}(0, \phi(0))(\theta'(0))^3\right\}.
\end{aligned}$$

Here all functions satisfy the conditions of Theorem 4.1.3 and E is the region shown at the beginning of this section and in Theorems 4.1.2 and 4.1.3.

We chose $N = 3$ and order number $n = 10$ for Simpson's composite rule because 3 and 10 are relatively prime. Thus, the numerator of the fraction $\langle Nx_i \rangle$ (where $x_i = \frac{i}{10}$, $i = 0, 1, \ldots, 9$) takes all elements of the complete residue system of mod 10. Similarly, we can take $N = 3$ and $n = 14$ and $N = 5$ and $n = 16$ (see Example 4.1.3.) etc.

With a loss of some accuracy, we can eliminate the last two terms, the ones with f''_{yy}, in the approximation expression and obtain the numerical quadrature formula

$$\iint_E f(x, y)dS \approx \frac{1}{15}\sum_{j=1}^{9}\frac{4}{3 + (-1)^j}\Phi\left(\frac{j}{10}, \langle\frac{3j}{10}\rangle\right).$$

Obviously, the above formula can also be established directly from Theorem 4.1.2.

Example 4.1.3. Take $N = 5$ and order number $n = 16$ for Simpson's composite rule. From Theorem 4.1.3,

$$\begin{aligned}
&\iint_E f(x, y)dS \\
\approx\; & \tfrac{1}{6}\textstyle\sum_{j=1}^{15}\tfrac{1}{3+(-1)^j}\Phi\left(\tfrac{j}{16}, \langle\tfrac{5j}{16}\rangle\right) \\
& +\frac{1}{3150000}\left[f''_{yy}(1, \phi(1))(\theta'(1))^3 - f''_{yy}(0, \phi(0))(\theta'(0))^3\right].
\end{aligned}$$

The above two numerical quadrature formulas are simple. If we need formulas with more accuracy, we can use Theorem 4.1.2 or Theorem 4.1.3 with larger N and n.

4.2 Numerical integration of oscillatory integrals

For large real parameter λ, the integral on the left-hand side of equation (3.4.9) in the basic expansion theorem 3.4.1 is an oscillatory integral. Obviously, traditional numerical integration methods cannot be applied on the oscillatory integrals. However, equation (3.4.9) gives a method for evaluating the oscillatory integrals approximately because all of the integrals on the right-hand side of equation (3.4.9) are no longer oscillatory integrals. In fact, for $\lambda > 10$ or $\Lambda > 15$, only a few terms on the right-hand side of equation (3.4.9) are needed to get very good approximations of the oscillatory integral. In particular, formula (3.4.9) is suitable for a nonseparable integrand $f(x, y)$ (i.e., $f(x, y)$ is not necessarily a function with the form $f(x)g(y)$).

We now consider numerical integration of oscillatory integrals with several large real parameters.

Assume that $\Phi(x, y_1, \dots, y_k)$ is a continuous function defined on the domain $D = \{(x, y) : a \leq x \leq b, -\infty < y_i < \infty, i = 1, 2, \dots, k)$. In addition, $f(\dots, y_i, \dots)$ is a periodic function of y_i with period 1. In this section, we will discuss numerical integration of the oscillatory integrals

$$I = \int_a^b \Phi(x, N_1x, N_2x, \dots, N_kx)dx$$

and

$$J = \int_V f(x_1, \dots, x_s, N_1x_1, \dots, N_kx_k)dV,$$

where V is a higher dimensional cube with the dimension $\max(s, k)$; dV is the volume measure; and $N_1, \cdots, N_k$ are infinitely large real parameters with different orders or arbitrarily large parameters. For the sake of simplicity, we will denote $(x_1, \dots, x_s)$, $(y_1, \cdots, y_k)$, $dx_1 \cdots dx_s$, and $dy_1 \cdots dy_k$ by X, Y, dX, and dY, respectively.

By using the basic lemma 3.2.3, we obtain the following theorem.

Theorem 4.2.1 *Let $f(X, Y) = f(x_1, \dots, x_s; y_1, \dots, y_k)$ be a continuous function defined on $V_{s+k} = [0, 1]^{s+k}$ that has the period 1 in terms of each variable y_i $(i = 1, \dots, k)$. Denote $\nu = \max\{s, k\}$; $\mu = \min\{s, k\}$; $dX = dx_1, \dots, dx_s$; $dY = dy_1, \dots, dy_k$; and by dV the volume measure. Then*

$$\int_{V_\nu} f(x_1, \dots, x_s; N_1x_1, \dots, N_kx_k)dV = \int_{V_{s+k}} f(X, Y)dXdY + O\left[\omega\left(f; \frac{1}{N_1}, \dots, \frac{1}{N_\mu}, 0, \dots, 0\right)\right]. \tag{4.2.1}$$

Proof. We prove the theorem in two steps. First, we consider the case when $s \le k$. In this case, the integral on the right-hand side can be written as a $2k$-dimensional integral; i.e.,

$$\int_{V_{s+k}} f(X;Y)dXdY = \int_{V_{2k}} f(X;Y)dx_1\cdots dx_k dy_1\cdots dy_k. \tag{4.2.2}$$

Denote

$$\begin{aligned} f(X;Y) &= f(x_1,\dots,x_s;y_1,\dots,y_k)(x_{s+1}\cdots x_k)^0 \\ &= g(x_1,\dots,x_k;y_1,\dots,y_k). \end{aligned} \tag{4.2.3}$$

Thus,

$$\begin{aligned} &\omega\left(g;\tfrac{1}{N_1},\dots,\tfrac{1}{N_k};0,\dots,0\right) \\ =\ &\omega\left(f;\tfrac{1}{N_1},\dots,\tfrac{1}{N_s};0,\dots,0\right). \end{aligned}$$

Applying the basic lemma 3.2.3 to the right-hand side of equation (4.2.2) yields

$$\begin{aligned} &\int_{V_{2k}} f(X;Y)dV \\ =\ &\int_{V_k} f(x_1,\dots,x_s;N_1x_1,\dots,N_kx_k)dV \\ &+O\left[\omega\left(f;\frac{1}{N_1},\dots,\frac{1}{N_s};0,\dots,0\right)\right]. \end{aligned} \tag{4.2.4}$$

Since $k = \nu = \max\{s,k\}$ and $s = \mu = \min\{s,k\}$, when $s \le k$, we know equation (4.2.1) to be true from (4.2.2) and (4.2.4).

For the case of $s > k$, we can rewrite the integral on the right-hand side of equation (4.2.1) as

$$\begin{aligned} \int_{V_{s+k}} f(X;Y)dXdY &= \int_{V_{s-k}} dx_{k+1}\cdots dx_s \\ &\quad\times \int_{V_{2k}} f(X;Y)dx_1\cdots dx_k dy_1\cdots dy_k. \end{aligned} \tag{4.2.5}$$

Consider $x_{k+1},\dots,x_s$ as fixed parameters, then the above integrand can be denoted

$$f(X;Y) = g(x_1,\dots,x_k;y_1,\dots,y_k).$$

Therefore,

$$\begin{aligned} &\omega\left(g;\tfrac{1}{N_1},\dots,\tfrac{1}{N_k};0,\dots,0\right) \\ =\ &\omega\left(f;\tfrac{1}{N_1},\dots,\tfrac{1}{N_k};0,\dots,0\right). \end{aligned}$$

Hence, an application of the basic lemma 3.2.3 to the right-hand integral of equation (4.2.5) yields

$$\begin{aligned}
&\int_{V_{s+k}} f(X;Y)dXdY \\
= &\int_{V_{s-k}} dx_{k+1}\cdots dx_s \int_{V_k} f(X; N_1x_1, \dots, N_kx_k)dx_1\cdots dx_k \\
&+\int_{V_{s-k}} O\left[\omega\left(f; \frac{1}{N_1}, \dots, \frac{1}{N_k}; 0, \dots, 0\right)\right] dx_{k+1}\cdots dx_s \\
= &\int_{V_s} f(X; N_1x_1, \dots, N_kx_k)dV \\
&+O\left[\omega\left(f; \frac{1}{N_1}, \dots, \frac{1}{N_k}; 0, \dots, 0\right)\right].
\end{aligned}$$

Because $s = \nu = \max\{s, k\}$ and $k = \mu = \min\{s, k\}$, the above equation shows that equation (4.2.1) is also true for the case of $s > k$. Hence, we have proved that the theorem holds for all cases. □

From equation (4.2.1), it is easy to derive the following corollary.

Corollary 4.2.2 *Preserve the conditions of Theorem 4.2.1, and further assume that $f(X;Y)$ has continuous derivatives with respect to each x_i $(i = 1, \dots, k)$. We then have*

$$\begin{aligned}
&\left|\int_{V_\nu} f(x_1, \dots, x_s; N_1x_1, \dots, N_kx_k)dV - \int_{V_{s+k}} f(X;Y)dV\right| \\
\leq\ &A\left(\frac{1}{N_1} + \cdots + \frac{1}{N_\mu}\right),
\end{aligned} \tag{4.2.6}$$

where A is a constant, $\nu = \max\{s, k\}$, and $\mu = \min\{s, k\}$.

In general, it is hard to evaluate oscillatory integrals. However, dimen-sionality-reducing formula (4.2.1) and estimation (4.2.6) give a method for evaluating multivariate oscillatory integrals with large real parameters $N_1, \cdots, N_k$ by using regular multivariate integrals. In addition, the approximate error is of the same order as $\omega\left(f; \frac{1}{N_1}, \cdots, \frac{1}{N_\mu}; 0, \dots, 0\right)$ or $\frac{1}{N_1} + \cdots + \frac{1}{N_\mu}$. Hence, the larger the values taken by the parameters, the better the approximation given by formula (4.2.1).

Repeating Theorem 4.2.1, we find an interesting result. Assume that $f(x_1, y_1, \dots, y_k)$ is a periodic continuous function with respect to each y_i $(i = 1, \dots, k)$. We replace y_1 in the integral

$$I = \int_{V_{k+1}} f(x_1; y_1, \dots, y_k)dx_1dy_1\cdots dy_k$$

by N_1x_1, then from dimensionality-reducing formula (4.2.1) we obtain

$$\begin{aligned} I &= \int_{V_k} f(x_1; N_1x_1, y_2, \ldots, y_k)dx_1dy_2\cdots, dy_k \\ &\quad + O\left[\omega\left(f; \tfrac{1}{N_1}, 0, \ldots, 0\right)\right]. \end{aligned}$$

Similarly, we replace y_2 by $N_1N_2x_1$ in the right-hand integral of the above equation. We also note that for the function

$$f(x_1; N_1x_1, y_2, \ldots, y_k) = g(x_1, y_2, \ldots, y_k),$$

there exists

$$\begin{aligned} \omega\left(g; \tfrac{1}{N_1N_2}, 0, \ldots, 0\right) &= O\left[\omega\left(f; \tfrac{1}{N_1N_2}, \tfrac{1}{N_2}, 0, \ldots, 0\right)\right] \\ &= O\left[\omega\left(f; \tfrac{1}{N_1}, \tfrac{1}{N_2}, 0, \ldots, 0\right)\right]. \end{aligned}$$

Dimensionality-reducing formula (4.2.1) gives

$$\begin{aligned} I &= \int_{V_{k-1}} f(x_1; N_1x_1, N_1N_2x_1, y_3, \ldots, y_k)dx_1dy_3\cdots dy_k \\ &\quad + O\left[\omega\left(f; \tfrac{1}{N_1}, \tfrac{1}{N_2}, 0, \cdots, 0\right)\right]. \end{aligned}$$

We continue this process until all y_i are replaced and in the end obtain

$$\begin{aligned} I &= \int_0^1 f(x_1; N_1x_1, N_1N_2x_1, \ldots, N_1 \ldots N_kx_1)dx_1 \\ &\quad + O\left[\omega\left(f; \tfrac{1}{N_1}, \tfrac{1}{N_2}, \ldots, \tfrac{1}{N_k}, 0\right)\right]. \end{aligned}$$

Denote $\lambda_1 = N_1$, $\lambda_i = N_1N_2\cdots N_i$ $(1 \leq i \leq k)$; the following theorem summarizes these results.

Theorem 4.2.3 *Let $f(x, y_1, \ldots, y_k)$ be a continuous function defined on V_{k+1} and $f(x, \ldots, y_i, \ldots)$ $(i = 1, \ldots, k)$ a periodic function with period 1. Suppose $\{\lambda_i\}_{i=1}^k$ is a set of infinitely large variables that satisfies $\lambda_1 \to \infty$, $\lambda_i/\lambda_{i-1} \to \infty$ $(i = 2, 3, \ldots, k)$. Then we have the dimensionality-reducing formula*

$$\begin{aligned} &\int_0^1 f(x, \lambda_1x, \lambda_2x, \ldots, \lambda_kx)dx \\ &= \int_{V_{k+1}} f(x, y_1, y_2, \ldots, y_k)dV \\ &\quad + O\left[\omega\left(f; \frac{1}{\lambda_1}, \frac{\lambda_1}{\lambda_2}, \ldots, \frac{\lambda_{k-1}}{\lambda_k}, 0\right)\right]. \end{aligned} \tag{4.2.7}$$

Obviously, the method in Theorem 4.2.3 can be used to solve the problems raised at the beginning of this section. Let us assume that in formula (4.2.7) $\{\lambda_i\}_{i=1}^k$ is a set of infinitely large variables; i.e., $1 << \lambda_1 << \lambda_2 << \cdots << \lambda_k$ (λ_k is much larger than λ_{k-1}, λ_{k-1} is much larger than λ_{k-2}, ..., λ_2 is much larger than λ_1, and λ_1 is much larger than 1). Then, we can drop the remainder and obtain an approximation of the oscillatory integral as follows.

$$\int_0^1 f(x, \lambda_1 x, \ldots, \lambda_k x)dx \approx \int_{V_{k+1}} f(x, y_1, \ldots, y_k)dV. \qquad (4.2.8)$$

Obviously, evaluation of the right-hand side regular multivariate integral is much easier than that of the oscillatory integral itself. We now consider two examples.

Example 4.2.1 Let $f(x)$ be a continuous function defined on $[a, b]$, and let θ, N_1, N_2 be real numbers satisfying $\theta > 0$, $N_2 >> N_1 >> |b - a|$. Then

$$\int_a^b f(x) \cos(2\theta \sin(N_1 x) \sin(N_2 x))dx$$
$$\approx (J_0(\theta))^2 \int_a^b f(x)dx, \qquad (4.2.9)$$

where $J_0(\theta)$ is the Bessel function of order zero. In fact, according to formula (4.2.7) or (4.2.8), the left-hand integral of (4.2.9) is approximately equal to

$$\left(\frac{1}{2\pi}\right)^2 \int_0^{2\pi} \int_0^{2\pi} \cos(2\theta \sin y_1 \sin y_2)dy_1 dy_2 \int_a^b f(x)dx$$
$$\approx (J_0(\theta))^2 \int_a^b f(x)dx.$$

Example 4.2.2 Let $f(x) \in C([a, b])$, $N_2 >> N_1 >> |b - a|$, and $\alpha^2 + \beta^2 = 1$. Then because

$$\int_0^{\pi/2} \int_0^{\pi/2} \frac{(\alpha^2 \cos^2 \theta + \beta^2 \cos^2 \phi)d\theta d\phi}{\sqrt{(1 - a^2 \sin^2 \theta)(1 - \beta^2 \sin^2 \phi)}} = \frac{\pi}{2}$$

formula (4.2.8) gives the approximation

$$\int_a^b f(x) \frac{a^2 \cos^2(N_1 x) + \beta^2 \cos^2(N_2 x)}{\sqrt{(1 - a^2 \sin^2 N_1 x)(1 - \beta^2 \sin^2 N_2 x)}} dx$$
$$\approx \frac{2}{\pi} \int_a^b f(x)dx.$$

The above two examples show that an oscillatory integral with several large parameters can sometimes be approximated by a single integral. Obviously, the direct computation of the oscillatory integrals are very difficult.

4.3 Numerical quadrature of strongly oscillatory integrals with compound precision

Various numerical integration methods have been developed for the Fourier type of oscillatory integrals and a comprehensive exposition can be found in Davis and Rabinowitz's book [19]. In particular, Havie [26] has shown that an Euler–Maclaurin type expansion formula could be used as an effective tool to treat oscillatory integrals (see also [37]). The object of this section is to construct a class of quadrature formulas that can have any given degree of compound precision for integrals of the form shown in (3.4.1):

$$\bar{I}(f;\lambda) := \int_0^1 f(x, \lambda x)dx, \tag{4.3.1}$$

where $f(\cdot, y)$ is a periodic function of y with period 1, and λ is a real parameter that may take very large values that are more than 100. Similar to the notations used in Section 3.4, throughout this section we assume that

$$f_x^{(j)}(x, y) \equiv \partial^j f / \partial x^j \in C(V_2), \tag{4.3.2}$$

where $j \geq 2$, $V_2 \equiv [0, 1]^2$, and $C(V_2)$ denotes the set of continuous functions defined on V_2.

For given positive integers p, q, and r, we say that a numerical integration formula $\bar{I}(f;\lambda) \approx S(f;\lambda)$ possesses $[p, q, \lambda^{-r}]$ *degree of compound precision* if for any algebraic-trigonometric polynomial in x and y of degree not exceeding p and q, respectively, of the form

$$P(x, y) = \sum_{\substack{0 \leq \nu \leq p \\ 0 \leq \mu \leq q}} \left[c_{\nu\mu} x^\nu \cos(2\pi \mu y) + d_{\nu\mu} x^\nu \sin(2\pi \mu y)\right], \tag{4.3.3}$$

$c_{\nu\mu}$, $d_{\nu\mu}$ being arbitrary constants, there exists a constant $K > 0$, independent of λ, such that

$$|\bar{I}(P;\lambda) - S((P;\lambda)| < K\lambda^{-m} \tag{4.3.4}$$

holds for all sufficiently large λ (say, $\lambda \geq 100$).

Generally, for a polynomial of the form (4.3.3) the constant K depends only upon m and the norm $\| P \| = \max_{V_2} |P(x, y)|$, or only on m and the norms of derivatives $P_x^{(k)}(x, y)$, $(k = m, m-1)$. An explicit estimate of the constant will be given near the end of the section.

To construct the numerical integration formula $\bar{I}(f;\lambda) \approx S(f;\lambda)$, first we use the basic expansion theorem of integrals with parameters 3.4.1 to give an approximate representation of $\bar{I}(f;\lambda)$, which is defined by (4.3.1). We will show that $\bar{I}(f;\lambda)$ may be represented approximately by some easily computed integrals with an error term $O(\lambda^{-m})$, $(\lambda \to \infty)$. For large λ, the constant implied by $O(\cdot)$ may be evaluated explicitly.

Theorem 4.3.1 *For large λ, we have the following expansion of the $\bar{I}(f;\lambda)$ defined by (4.3.1).*

$$\bar{I}(f;\lambda) = \iint_{V_2} f(x,y)dxdy + \sum_{j=1}^{r-1} \frac{\lambda^{-j}}{j!} \int_0^1 J(f;y,j,\lambda)dy + \rho_r, \tag{4.3.5}$$

where $J(f;y,j,\lambda)$ is defined by

$$J(f;y,j,\lambda) = \left[f_x^{(j-1)}(t,y)\bar{B}_j(y-\lambda t) \right]_{t=0}^{t=1} \tag{4.3.6}$$

and the remainder ρ_r can be represented by

$$\rho_r = \frac{1}{r!}\lambda^{-r} \left\{ \int_0^1 J(f;y,r,\lambda)dy - \int_0^1 dx \int_0^1 \bar{B}_r(y-\lambda x) f_x^{(r)}(x,y)dy \right\}. \tag{4.3.7}$$

In the above representation, some notations defined in Section 3.4 *are used: $\bar{B}_j(t)$ is the Bernoulli function of period* 1 *that coincides with the jth degree Bernoulli polynomial $B_j(t)$ on* $[0,1]$ *and* $[\psi(t)]_{t=a}^{t=b} = \psi(b) - \psi(a)$.

Proof. The theorem can be considered as a modification of the basic expansion theorem (Theorem 3.4.2), which can be verified by using the Euler–Maclaurin summation formula of the form

$$\begin{aligned}\sum_{a \le k < b} g(k) &= \int_a^b g(x)dx + \sum_{j=1}^{r} \frac{1}{j!} \left[g^{(j-1)}(t)\bar{B}_j(-t) \right]_a^b \\ &\quad - \frac{1}{r!} \int_a^b \bar{B}_r(-x)dg^{(r-1)}(x),\end{aligned} \tag{4.3.8}$$

which is valid for any $g(x)$ such that $g^{(r-1)}(x)$ is of bounded variation on $[a,b]$. Here we need only the case where $g^{(r)}(x)$ is continuous.

By substituting $y = \lambda x$ and noting the periodicity of function f we have

$$\begin{aligned}\bar{I}(f;\lambda) &= \frac{1}{\lambda} \int_0^\lambda f\left(\frac{y}{\lambda}, \langle y \rangle\right) dy \\ &= \int_0^1 \frac{1}{\lambda} \sum_{-y \le k < \lambda - y} f\left(\frac{k+y}{\lambda}, y\right) dy.\end{aligned} \tag{4.3.9}$$

Thus an application of (4.3.8) to the inner summation of (4.3.9) (with $a = -y$ and $b = \lambda - y$) yields the following

$$\begin{aligned}
&\frac{1}{\lambda} \sum_{-y \le k < \lambda - y} f\left(\frac{k+y}{\lambda}, y\right) \\
= &\frac{1}{\lambda} \int_{-y}^{\lambda - y} f\left(\frac{u+y}{\lambda}, y\right) du \\
&+ \sum_{j=1}^{r} \frac{\lambda^{-j}}{j!} \left[f_x^{(j-1)}(1, y) \bar{B}_j(y - \lambda) - f_x^{(j-1)}(0, y) \bar{B}_j(y) \right] \\
&- \frac{\lambda^{-r}}{r!} \int_{-y}^{\lambda - y} \bar{B}_r(-u) d_u f_x^{(r-1)} \left(\frac{u+y}{\lambda}, y\right) \\
= &\int_0^1 f(x, y) dx + \sum_{j=1}^{r} \frac{\lambda^{-j}}{j!} J(f; y, j, \lambda) \\
&- \frac{\lambda^{-r}}{r!} \int_0^1 f_x^{(r)}(x, y) \bar{B}_r(y - \lambda x) dx.
\end{aligned}$$

Hence the theorem is justified by means of the last equation and (4.3.9). □

Let us denote

$$\| B_r \| = \max_{0 \le x \le 1} |B_r(x)|, \quad \| f_x^{(k)} \| = \max_{V_2} |f_x^{(k)}(x, y)|.$$

It is easy to get an estimate of $|\rho_r|$ as follows:

$$|\rho_r| \le \frac{1}{r! \lambda^r} \left(2 \| f_x^{(r-1)} \| + \| f_x^{(r)} \| \right) \| B_j \| . \tag{4.3.10}$$

In order to obtain quadrature formulas using only integrand function values, it is necessary to approximate the partial derivatives contained in (4.3.6). This may be accomplished by using a pair of numerical differentiation formulas of the form

$$h^j f^{(j)}(0) = \sum_{k=0}^{r-1} C_k^{(j)} f(kh) + E_1^j h^r \tag{4.3.11}$$

and

$$h^j f^{(j)}(1) = \sum_{k=0}^{r-1} D_k^{(j)} f(1 - kh) + E_2^j h^r. \tag{4.3.12}$$

Assume that $f^{(r)}(t) \in C([0, 1])$, $(r-1)h \le 1$, $0 \le k \le r-1$, and let $(t)_r := t(t+1) \cdots (t+r-1)$. Then for convenience we may simply make use of Markov's formula

$$\begin{aligned}
h^j f^{(j)}(0) = &\sum_{k=j}^{r-1} (-1)^{k-j} \frac{1}{k!} \Delta_h^k f(0) \left[\left(\frac{d}{dt}\right)^j (t)_k \right]_{t=0} \\
&+ (-1)^{r-j} h^r \tfrac{1}{r!} f^{(r)}(\xi) \left[\left(\tfrac{d}{dt}\right)^j (t)_r \right]_{t=0},
\end{aligned}$$

$\xi \in [0, (r-1)h]$, which may be rewritten in the form (4.3.11) with the remainder factor

$$\begin{aligned} E_1^j &= \frac{(-1)^{r-j}}{r!} f^{(r)}(\xi) \left[\left(\frac{d}{dt} \right)^j (t)_r \right]_{t=0} \\ &= \frac{(-1)^{r-j}}{r!} f^{(r)}(\xi) |S_1(r, j)| j!, \end{aligned}$$

where $S_1(r, j)$ is the Stirling number of the first kind defined by

$$t(t-1)\cdots(t-r+1) = \sum_{k=1}^{r} S_1(r, k) t^k, \quad r \geq 1.$$

Consequently we have an estimate for $|E_1^j|$:

$$|E_1^j| \leq \frac{j!}{r!} \parallel f^{(r)} \parallel \cdot |S_1(r, j)|,$$

where $\parallel f^{(r)} \parallel = \max_{0 \leq t \leq 1} |f^{(r)}(t)|$. Similarly, with h being replaced by $-h$, and assuming $(r-1)h \leq 1$, we get another Markov formula (4.3.12) in which $|E_2^j|$ has the same estimate

$$|E_2^j| \leq \frac{j!}{r!} \parallel f^{(r)} \parallel \cdot |S_1(r, j)|.$$

Now for each fixed $y \in [0, 1]$ and with $h = \lambda^{-\alpha}$ $(0 < \alpha \leq 1)$, applying (4.3.11) and (4.3.12) to $f_x^{(j)}(t, y)$ at $t = 0$ and $t = 1$ respectively, one can obtain an approximation $\tilde{J}(f; y, j+1, \lambda)$ of $J(f; y, j+1, \lambda)$, $1 \leq j \leq r-2$, namely

$$\begin{aligned} \tilde{J}(f; y, j+1, \lambda) &= \lambda^{\alpha j} \bar{B}_{j+1}(y - \lambda) \sum_{k=0}^{r-1} B_k^{(j)} f(1 - k\lambda^{-\alpha}, y) \\ &\quad - \lambda^{\alpha j} \bar{B}_{j+1}(y) \sum_{k=0}^{r-1} A_k^{(j)} f(k\lambda^{-\alpha}, y). \end{aligned} \tag{4.3.13}$$

In particular, we define

$$\tilde{J}(f; y, 1, \lambda) \equiv J(f; y, 1, \lambda) = \left[f(t, y) \bar{B}_1(y - \lambda t) \right]_{t=0}^{t=1}. \tag{4.3.14}$$

Then replacing $J(f; y, j, \lambda)$ of (4.3.5) by (4.3.13)–(4.3.14), we get an approximation for $\bar{I}(f; \lambda)$,

$$\tilde{I}(f; \lambda) \equiv \iint_{V_2} f(x, y) dx dy + \sum_{j=0}^{r-2} \frac{\lambda^{-j-1}}{(j+1)!} \int_0^1 \tilde{J}(f; y, j+1, \lambda) dy. \tag{4.3.15}$$

□

Theorem 4.3.2 *For a given integer $m \geq 3$ and a real number α with $(r-2)/(r-1) \leq \alpha \leq 1$, the error term involved in $\tilde{I}(f;\lambda)$ is of the order $O(\lambda^{-r})$ $(\lambda \to \infty)$, provided that $f_x^{(r)}(x,y) \in C(V_2)$. More precisely, we have*

$$|\bar{I}(f;\lambda) - \tilde{I}(f;\lambda)| \leq \frac{\lambda^{-r}}{r!}\left[\sum_{j=2}^{r} \gamma_j \parallel B_j \parallel\right], \tag{4.3.16}$$

where γ_j is given by

$$\gamma_j = \begin{cases} (2/j)|S_1(r,j-1)| \cdot \parallel f_x^{(r)} \parallel, & \text{if } 2 \leq j \leq r-1, \\ 2 \parallel f_x^{(r-1)} \parallel + \parallel f_x^{(r)} \parallel, & \text{if } j = r. \end{cases} \tag{4.3.17}$$

Proof. By (4.3.5), (4.3.6), (4.3.11), (4.3.12), and (4.3.14), we have

$$\begin{aligned} & |\bar{I}(f;\lambda) - \tilde{I}(f;\lambda)| \\ \leq \ & |\rho_r| + \sum_{j=1}^{r-2} \frac{\lambda^{-j-1}}{(j+1)!} \int_0^1 |\tilde{J}(f;y,j+1,\lambda) - J(f;y,j+1,\lambda)| dy \\ \leq \ & |\rho_r| + \sum_{j=1}^{r-2} \frac{\lambda^{-j-1}}{(j+1)!} \parallel B_{j+1} \parallel \left(|E_1^j| + |E_2^j|\right) h^{r-j}. \end{aligned}$$

Here both E_1^j and E_2^j include a parameter $y \in [0,1]$. However, they can be estimated by

$$\max\left\{|E_1^j|, |E_2^j|\right\} \leq \frac{j!}{r!} \parallel f_x^{(r)}(x,y) \parallel \cdot |S_1(r,j)|.$$

Recall that $h = (1/\lambda)^\alpha$, $((r-2)/(r-1) \leq \alpha \leq 1)$, so for large λ,

$$\begin{aligned} \lambda^{-j-1} h^{r-j} &= (1/\lambda)^{\alpha(r-j)+j+1} \\ &\leq (1/\lambda)^{\left(\frac{r-j-1}{r-j}\right)(r-j)+j+1} = \lambda^{-r}. \end{aligned}$$

Consequently

$$|\bar{I}(f;\lambda) - \tilde{I}(f;\lambda)| \leq |\rho_r| + \frac{\lambda^{-r}}{r!} \sum_{j=1}^{r-2} \frac{2}{j+1} \parallel B_{j+1} \parallel \cdot \parallel f_x^{(r)} \parallel \cdot |S_1(r,j)|.$$

Hence (4.3.16) is obtained with the aid of (4.3.10). □

We now discuss the construction of quadrature formulas with compound precision of $\bar{I}(f;\lambda)$ from the approximation representations shown in Theorems 4.3.1 and 4.3.2.

Take an interpolatory quadrature formula (or Gaussian formula) with degree p of algebraic precision for an integral on [0, 1],

$$\int_0^1 \psi(x)dx \approx \sum_\mu a_\mu \psi(x_\mu), \quad x_\mu \in [0, 1].$$

Also take a formula of degree q of trigonometric precision and with equidistant knots and equal coefficients for an integral of a function with period unity,

$$\int_0^1 \phi(y)dy \approx \sum_{\nu=1}^{q+1} b_\nu \phi(y_\nu), \quad b_\nu = \frac{1}{q+1}, \; y_\nu = \frac{\nu}{q+1}.$$

Then the product formula

$$\iint_{V_2} f(x, y)dxdy \approx \sum_\mu \sum_\nu a_\mu b_\nu f(x_\mu, y_\nu) \tag{4.3.18}$$

has degree of algebraic-trigonometric precision (p, q) (see [25]).

Notice that $\tilde{J}(f; y, \cdot, \cdot)$ defined by (4.3.13) and (4.3.14) are periodic functions of period 1 in y. We have worked with integrals of the form

$$J_j = \int_0^1 \bar{B}_j(y - \theta) f(\zeta, y)dy, \tag{4.3.19}$$

which may be rewritten as

$$J_j = \int_{-\theta}^{1-\theta} \bar{B}_j(t) f(\zeta, t + \theta)dt = \int_0^1 B_j(t) f(\zeta, t + \theta)dt, \tag{4.3.20}$$

where $\theta = 0$ or $\theta = \lambda$; $\xi = 0$ or $\xi = 1$ for $j = 1$; and $\xi = \mu\lambda^{-\alpha}$ or $\xi = 1 - \mu\lambda^{-\alpha}$ for $j = 2, \dots, r - 1$.

For a given weight function $B_j(x)$, $1 \le j \le r - 1$, let us now construct a formula with trigonometric precision q and of the form

$$\int_0^1 B_k(x)\phi(x)dx \approx \sum_{\nu=1}^{N} C_\nu^{(k)} \phi(x_\nu), \quad N = 2q + 1, \tag{4.3.21}$$

where $\phi(x)$ is a periodic function of period 1, and $0 \le x_1 < x_2 < \cdots < x_N \le 1$ is a given partition of [0, 1]. The degree of precision, q, requires (4.3.21) to be exact for all trigonometric polynomials $\phi(x) := T(x)$ of the form

$$T(x) = c_0 + \sum_{j=1}^{q} \left[c_j \cos(2\pi jx) + d_j \sin(2\pi jx)\right].$$

Consequently we get a system of linear equations

$$\begin{cases} \sum_{\nu=1}^{N} C_\nu^{(k)} = \int_0^1 B_k(x)dx = 0, \\ \sum_{\nu=1}^{N} C_\nu^{(k)} \cos(2\pi j x_\nu) = \xi_{kj}, \\ \sum_{\nu=1}^{N} C_\nu^{(k)} \sin(2\pi j x_\nu) = \eta_{kj}, \end{cases} \tag{4.3.22}$$

where $j = 1, 2, \ldots, q$ ($q = (N-1)/2$). Here, ξ_{kj} and η_{kj} are given by

$$\xi_{kj} = \int_0^1 B_k(x) \cos(2\pi j x)dx = -(-1)^{k/2} k!/(2\pi j)^k \tag{4.3.23}$$

and $\eta_{kj} = 0$ for even k, while

$$\eta_{kj} = \int_0^1 B_k(x) \sin(2\pi j x)dx = -(-1)^{(k-1)/2} k!/(2\pi j)^k \tag{4.3.24}$$

and $\xi_{kj} = 0$ for odd k.

Let $M \equiv M(x_1, \ldots, x_N)$ denote the coefficient matrix of the linear system (4.3.22). In what follows we always take $x_\nu = \nu/N = \nu/(2q+1)$, $\nu = 1, \ldots, N$. Certainly $M(1/N, 2/N, \ldots, N/N)$ is nonsingular. In fact, its corresponding determinant, det M, has the value

$$det\ M = 2^{-N}(-1)^{q(q+2)/2} \Pi_{0 \le \nu < \mu \le 2q} \left(\xi^\nu - \xi^\mu\right), \tag{4.3.25}$$

where $\xi = e^{2\pi i/N}$, $i = \sqrt{-1}$, is the Nth root of unity with $N = 2q+1$, and the product is taken over all integer pairs (ν, μ) within the interval $[0, 2q]$.

Thus it is clear that a quadrature formula of any given degree of trigonometric precision and of the form

$$\int_0^1 B_j(t)\phi(t)dt \approx \sum_{\nu=1}^{N} C_\nu^{(j)} \phi\left(\frac{\nu}{N}\right), \tag{4.3.26}$$

where $N = 2q+1$, can always be constructed explicitly.

Let us now denote

$$\sigma(f, \lambda, 0) = \sum_{\nu=1}^{N} C_\nu^{(1)} \left[f\left(1, \frac{\nu}{N} + \lambda\right) - f\left(0, \frac{\nu}{N}\right)\right], \tag{4.3.27}$$

and for $j \ge 1$,

$$\begin{aligned} \sigma(f, \lambda, j) &= \sum_{\mu=0}^{r-1} B_\mu^{(j)} \sum_{\nu=1}^{N} C_\nu^{(j+1)} f\left(1 - \mu\lambda^{-\alpha}, \frac{\nu}{N} + \lambda\right) \\ &\quad - \sum_{\mu=0}^{r-1} A_\mu^{(j)} \sum_{\nu=1}^{N} C_\nu^{(j+1)} f\left(\mu\lambda^{-\alpha}, \frac{\nu}{N}\right). \end{aligned} \tag{4.3.28}$$

Thus, according to (4.3.13)–(4.3.15) and (4.3.19)–(4.3.20), and making use of (4.3.26) we obtain

$$\int_0^1 \tilde{J}(f; y, j+1, \lambda)dy \approx \lambda^{\alpha j}\sigma(f, \lambda, j), \tag{4.3.29}$$

where $j = 0, 1, \ldots, r-2$.

Finally, substituting (4.3.29) into (4.3.15) and using (4.3.18), we arrive at the desired result $\bar{I}(f; \lambda) \approx S(f; \lambda)$ with

$$S(f; \lambda) = \sum_{\mu}\sum_{\nu} a_\mu b_\nu f(x_\mu, y_\nu) + \sum_{j=0}^{r-2} \frac{\lambda^{(\alpha-1)j-1}}{(j+1)!}\sigma(f, \lambda, j). \tag{4.3.30}$$

Here, a_μ and x_μ are chosen to satisfy formula (4.3.18); $b_\nu = 1/(q+1)$; $y_\nu = \nu/(q+1)$; $\nu = 1, \ldots, q+1$; $(r-2)/(r-1) \le \alpha \le 1$; and $N = 2q+1$.

Theorem 4.3.3 *For any positive integer p, q, and $r \ge 3$, and for any fixed parameter α with $(r-2)/(r-1) \le \alpha \le 1$, the numerical integration formula $\bar{I}(f; \lambda) \approx S(f; \lambda)$ always possesses the degree of compound precision $[p, q, \lambda^{-r}]$. More precisely, for all polynomials $P(x, y)$ of the form (4.3.3), the inequality*

$$|\bar{I}(P; \lambda) - S(P; \lambda)| \le \frac{\lambda^{-r}}{r!}\sum_{j=2}^{r} \gamma_j \parallel B_j \parallel, \tag{4.3.31}$$

where γ_j's are given by (4.3.17) with $f := P$, holds for all sufficiently large λ.

Proof. In view of (4.3.16), given by Theorem 4.3.2, it suffices to show that for $f(x, y) := P(x, y)$ we have

$$\tilde{I}(P; \lambda) = S(P; \lambda).$$

Let us recall (4.3.15) and note that

$$\iint_{V_2} P(x, y)dxdy = \sum_{\mu}\sum_{\nu} a_\mu b_\nu P(x_\mu, y_\nu).$$

In accordance with (4.3.30) it is sufficient to verify that (4.3.29) is exact for $f := P(x, y)$. More precisely, denote the expressions contained in (4.3.13)–(4.3.14) by $\tilde{J}(P; y, j+1, \lambda)$ and $\tilde{J}(P; y, 1, \lambda)$ respectively. Observe that all the $P(\cdot, y)$ as well as $P(\cdot, y+\theta)$, $\theta = 0$ or $\theta = \lambda$, are trigonometric polynomials in y with period 1, and of degrees not exceeding q. Thus from (4.3.13), (4.3.14), (4.3.19), (4.3.20), and (4.3.26) it follows that

$$\int_0^1 \tilde{J}(P; y, j+1, \lambda)dy = \lambda^{\alpha j}\sigma(P, \lambda, j),$$

where $j = 0, 1, \ldots, r-2$. This shows that (4.3.29) is exact for $f := P(x, y)$, and consequently (4.3.31) is proved. □

Notice that for $r \geq 2$ we have

$$\| B_r \| = \max_{0 \leq t \leq 1} |B_r(t)| \leq \frac{2(r!)}{(2\pi)^r} \xi(r),$$

where $\xi(r) = \sum_{k=1}^{\infty} (1/k)^r$, and

$$\sum_{k=1}^{r} |S_1(r, k)|(2\pi)^k = \Pi_{j=0}^{r-1} (1/2\pi + j) < \frac{r!}{2\pi}.$$

Utilizing these facts one may deduce from (4.3.31) the

Corollary 4.3.4 *There is an explicit estimation for the left-hand side of (4.3.31),*

$$|\bar{I}(P; \lambda) - S(P; \lambda)| < C\lambda^{-r}, \tag{4.3.32}$$

where C is given by

$$C = \left(\frac{(r-2)!}{6} + \frac{2\xi(r)}{(2\pi)^r} \right) \| P_x^{(r)} \| + \frac{4\xi(r)}{(2\pi)^r} \| P_x^{(r-1)} \| . \tag{4.3.33}$$

Remark 4.3.1. Clearly (4.3.30) involves a class of quadrature formulas, in which the parameter α may be chosen with some freedom, and all the constructive coefficients a_μ, $A_\mu^{(j)}$, $B_\mu^{(j)}$, and $C_\nu^{(j+1)}$ can be found by solving linear systems of equations with the aid of computers. Of course, one may also use other kinds of numerical differentiation formulas instead of Markov's.

Remark 4.3.2. A specialization of (4.3.30) to the case $f(x, y) = f(x)g(y)$ can be used to compute transformation integrals containing a large λ. For such a case the constructive process we have employed can be simplified.

4.4 Fast numerical computations of oscillatory integrals

In this section, we will describe Bradie, Coifman, and Grossmann's result [2] on treating a type of oscillatory boundary integral operator with or without singularities in $\mathbb{R}^2$. Also cursorily discussed is the extension to higher dimensions, which is straightforward.

The aim of treating the oscillatory integral operators is to solve the class of integral equations

$$\int_a^b k(s, t) f(t) dt = g(s), \tag{4.4.1}$$

where $f(t)$ is an unknown function but the kernel $k(s,t)$ and function $g(s)$ are given. The left-hand side of equation (4.4.1) can be considered as an integral operator on f. This type of integral equations, referred to as the *Fredholm, Volterra, or Weiner–Hopf integral equations*, appear frequently in *acoustic scattering problems, inverse problems*, and *electromagnetic scattering problems*. We will assume that equation (4.4.1) has a unique solution. In the *Galerkin method* for solving for the *weak solution* of an equation, the boundary of the domain of integration is approximated by discretizing it into several segments. The unknown function is then expanded in terms of certain known basis functions with unknown coefficients, which may occupy more than one segment. Finally, the resultant equation is tested with the same or different functions, resulting in a set of linear equations whose solution gives the unknown coefficients. Because of the nature of the integral operator, the matrix of the linear system is usually dense, and the inversion and final solution of the system is very time consuming, particularly for solving scattering problems that involve large objects.

In this section, we will use the results from [2] to show that, for certain boundary integral operators occurring in acoustic scattering computations, the matrix of the corresponding linear system is sparse when the unknown function is represented in the appropriate local cosine transform orthonormal basis. We will discuss two different integral operators, with the kernels $k(s,t) = e^{i\lambda|z(s)-z(t)|}$ and $e^{i\lambda|s-t|}/|s-t|$, where λ is a large real number greater than or equal to 100. More precisely, for the former kernel, if the closed curve Γ has the length L and a bounded curvature, then we have

$$H_\lambda(f)(s) = \int_0^L e^{i\lambda|z(s)-z(t)|} f(t)dt, \tag{4.4.2}$$

where s is the arc length parameter (i.e., $z(s) \in \Gamma$, $0 \le s \le L$). The numerical quadrature formulas of the above integral can be found in Section 3. This section will discuss the *sparsity of the matrix* of the linear system, which is from the Galerkin method and a discretization of integral (4.4.2). In order to discretize (4.4.2), we need $N \ge 2\lambda L$ discretization points, with respect to which we construct a quadrature formula of H. An equivalent method consists of the following two steps: (1) Find a finite orthogonal basis with N elements; (2) Examine the operator $H_N = P_N H P_N$, where P_N is the orthogonal projection on the span of the basis function.

Let $\{I_j = (a_j, a_{j+1})\}_{j=1}^M$ be a partition in Γ with $a_{M+1} = a_0$, which is always true for a closed boundary curve. Also let $\{b_j(s)\}_{j=1}^M$ be a set of window functions with the properties

(i) $\sum_{j=1}^M b_j^2(s) = 1$;

(ii) $b_j(s) = b_{j-1}(2a_j - s)$;

(iii) supp $(b_j(s)) \in I_{j-1} \cup I_j \cup I_{j+1}$ (supp (f) is the support of function f; i.e., the set on which f is not identically zero).

Denote by ds the arc measure. One can verify that the following functions form an orthonormal basis of $L^2(\Gamma, ds)$.

$$C_k^j(s) = \left(\frac{2}{a_{j+1} - a_j}\right)^{1/2} b_j(s) \cos\left(\left(k + \frac{1}{2}\right)\pi \frac{s - a_j}{a_{j+1} - a_j}\right). \tag{4.4.3}$$

For the truncation, we pick the points a_j to be equispaced with a suitable value of M and $a_{j+1} - a_j = L/\sqrt{N} = \eta$. We also choose $0 < k < \sqrt{N}$ and $b_j(s) = b\left(\frac{s-\alpha_j}{\eta}\right)$, where $\alpha_j = (a_j + a_{j+1})/2$ and $b(s)$ has its support in $[-3\eta/2, 3\eta/2]$ and satisfy $b(s) = b(-s)$ and $b^2(s)+b^2(s-\eta)+b^2(s+\eta) = 1$ for all $s \in [0, \eta/2]$. It is easy to see that all $b_j(s)$ satisfy conditions (i)–(iii) shown above. For instance,

$$\begin{aligned} & b_{j-1}(2a_j - s) = b\left(\tfrac{2a_j - s - \alpha_{j-1}}{\eta}\right) = b\left(\tfrac{\alpha_j - s}{\eta}\right) \\ = \; & b\left(\tfrac{s-\alpha_j}{\eta}\right) = b_j(s). \end{aligned}$$

Thus, we obtain the following discrete basis for Γ with N discrete equispaced points and each I_j having $\sqrt{N}$ points.

$$C_k^j(s) = \left(\frac{2}{\eta}\right)^{1/2} b\left(\frac{s - \alpha_j}{\eta}\right) \cos\left(\left(k + \frac{1}{2}\right)\Pi \frac{s - a_j}{\eta}\right), \tag{4.4.4}$$

where $0 \le j, k < \sqrt{N}$ and $\eta = L/\sqrt{N}$.

Let P_N denote the orthogonal projection on the span of the basis in (4.4.4); it converges to identity as N goes to infinity. In addition, if we want to limit the frequencies to be less than $\sqrt{N}$, we are forced to segment the curve Γ in no less than $\sqrt{N}$ windows.

As we mentioned at the beginning of the section, to apply the Galerkin method to solve the integral equation (4.4.1) with integral operator (4.4.2), we substitute an expansion of f in terms of the basis $C_k^j(s)$ with unknown coefficients $\mathbf{h} = \{h_{jk}\}$ into the integral equation. The resultant equation is tested with the basis $C_k^j(s)$, yielding a set of linear equations whose solution gives the unknown coefficients, as follows.

$$\left[A_{k,k'}^{j,j'}\right]\mathbf{h} = \langle g(s), C_{k'}^{j'}(s)\rangle, \tag{4.4.5}$$

where $A_{k,k'}^{j,j'} = \langle H_N C_k^j, C_{k'}^{j'}\rangle$, $H_N = P_N H_\lambda P_N$, and P_N is the orthogonal projection on the span of the basis functions. We now give an estimate of the sparsity of the matrix $\left[A_{k,k'}^{j,j'}\right]$.

Fix $I_j, I_{j'}, j \neq j'$, then

$$\begin{aligned} & A^{j,j'}_{k,k'} \\ = \quad & \langle H_N C^j_k, C^{j'}_{k'} \rangle \\ = \quad & \frac{4N}{L^2} \int_0^L \int_0^L e^{i\lambda |z(s)-z(t)|} b\left(\frac{t-\alpha_j}{\eta}\right) b\left(\frac{s-\alpha_{j'}}{\eta}\right) \\ & \times \cos\left[\left(k+\frac{1}{2}\right)\pi\frac{t-\alpha_j}{\eta}\right] \cos\left[\left(k'+\frac{1}{2}\right)\pi\frac{s-\alpha_{j'}}{\eta}\right] dsdt, \quad (4.4.6) \end{aligned}$$

$0 \leq k < \sqrt{N}, 0 \leq k' < \sqrt{N}$. We can find all k and k' such that

$$|A^{j,j'}_{k,k'}| < \epsilon,$$

for any given precision threshold ϵ.

A change of variables, $s - a_{j'} = \eta u$ and $t - a_j = \eta v$, yields

$$\begin{aligned} A^{j,j'}_{k,k'} \quad = \quad & 4 \int_{\mathbb{R}} \int_{\mathbb{R}} e^{i\lambda |z(a_{j'}+\eta u) - z(a_j + \eta v)|} b\left(v - \frac{1}{2}\right) b\left(u - \frac{1}{2}\right) \\ & \times \cos\left[\left(k+\frac{1}{2}\right)\pi v\right] \cos\left[\left(k'+\frac{1}{2}\right)\pi u\right] dudv. \quad (4.4.7) \end{aligned}$$

After representing $\cos(\cdot)$ in the integral of (4.4.7) as a combination of exponential functions, we estimate the resulting integral as

$$\int_{-1}^{1} \int_{-1}^{1} e^{i\lambda |z(a_{j'}+\eta u) - z(a_j+\eta v)|} \beta(v)\beta(u) e^{i\pm\pi(k'u+kv)} dudv,$$

where $\beta(u) = b\left(u - \frac{1}{2}\right) e^{\pm i\pi u/2}$. The integral can be written more compactly as

$$\int_{-1}^{1} \int_{-1}^{1} \beta(u)\beta(v) e^{i\Phi_N(u,v,k,k')} dudv, \quad (4.4.8)$$

where

$$\Phi_N(u, v, k, k') = \lambda |z(a_{j'} + \eta u) - z(a_j + \eta v)| \pm \pi(k'u + kv).$$

Hence,

$$\frac{\partial \Phi_N}{\partial u} = \lambda\eta \frac{z'(a_{j'} + \eta u) \cdot (z(a_{j'} + \eta u) - z(a_j + \eta v))}{|z(a_{j'} + \eta u) - z(a_j + \eta v)|} \pm \pi k' \quad (4.4.9)$$

and

$$\begin{aligned} & \frac{\partial \Phi_N}{\partial v} \\ = \quad & -\lambda\eta \frac{z'(a_j + \eta v) \cdot (z(a_{j'} + \eta u) - z(a_j + \eta v))}{|z(a_{j'} + \eta u) - z(a_j + \eta v)|} \pm \pi k. \quad (4.4.10) \end{aligned}$$

On the other hand,

$$\begin{aligned} z'(a_{j'} + \eta u) &= z'(a_{j'}) + O(\eta u) = z'(a_{j'}) + O\left(\frac{1}{\sqrt{N}}\right), \\ z'(a_j + \eta v) &= z'(a_j) + O\left(\frac{1}{\sqrt{N}}\right). \end{aligned} \tag{4.4.11}$$

Therefore, denoting

$$\Delta(u, v) = \frac{z(a_{j'} + \eta u) - z(a_j + \eta v)}{|z(a_{j'} + \eta u) - z(a_j + \eta v)|} \tag{4.4.12}$$

changes (4.4.9) and (4.4.10) to

$$\begin{aligned} \frac{\partial \Phi_N}{\partial u} &= \sqrt{\lambda}\left\{z'(a_{j'}) \cdot \Delta(u, v) \pm \pi \frac{k'}{\sqrt{N}}\right\} + O(1), \\ \frac{\partial \Phi_N}{\partial v} &= \sqrt{\lambda}\left\{-z'(a_j) \cdot \Delta(u, v) \pm \pi \frac{k}{\sqrt{N}}\right\} + O(1), \end{aligned} \tag{4.4.13}$$

where $O(1) \leq 1$.

$\Delta(u, v)$ describes a unit vector on an arc of the unit circle with a length of the order $1/r_{j,j'}$, $r_{j,j'}$ being equal to 1 plus the distance between the windows measured in units of η. Hence, the set $(z'(a_{j'}) \cdot \Delta(u, v) - Z'(a_j) \cdot \Delta(u, v))$ describes an arc of an ellipse with a length less than or equal to $c/r_{j,j'}$.

If the distance from $(\pm k/\sqrt{N}, \pm k'/\sqrt{N})$ to this arc is greater than $(\delta+1)/\sqrt{N}$, then

$$|\nabla \Phi_N(u, v)| = |\left(\frac{\partial \Phi_N}{\partial u}, \frac{\partial \Phi_N}{\partial v}\right)| \geq \delta.$$

(Note that $N \geq 2\lambda L$.) Applying integration by parts to integral (4.4.8) and using the above estimation, we find that integral (4.4.8) is dominated by c_j/δ^j for all j and constants c_j, which depend solely on the function $b(u)$. For an arbitrary, fixed, j we pick a δ such that $\frac{c_j}{\delta^j} < \epsilon$. Then in the square $[-1, 1]^2$, all the grid points, $(k/\sqrt{N}, k'/\sqrt{N})$, whose distance from the arc of the ellipse, with the length $1/r_{j,j'}$, exceeds $\delta/\sqrt{N}$ generate matrix entries less than ϵ. The number of grid points that does not meet this criteria is $\delta\sqrt{N}/r_{j,j'}$. With possibly a few exceptions, $r_{j,j'} \approx |j - j'| + 1$. Therefore, the total number of coefficients satisfying $|A^{j,j'}_{k,k'}| > \epsilon$ is approximately

$$C\delta\sqrt{N}\sqrt{N}\sum_{\ell=1}^{\sqrt{N}} \frac{1}{\ell} \leq C\delta N \log N.$$

If the curve has a corner or an irregular region with $\sqrt{N}$ discretization points and of the size $1/\sqrt{N}$, we can choose a window basis of $\sqrt{N}$ step functions that leads to an estimate of $|\partial\Phi_N/\partial u| > \delta$ for all k' windows outside an interval of length $1/r_{j,j'}$, giving $\delta(\sqrt{N}/r_{j_0,j'})\sqrt{N}$ as the count of the entries, (k, k'), for which $|A^{j,j'}_{k,k'}| \geq \epsilon$. Here, j_0 denotes the index of a bad window.

In a higher dimensional boundary setting, such as $\Gamma \in \mathbb{R}^2$, we make an arbitrary partition of nonoverlapping rectangles with sides parallel to the axes. Let the function defined on rectangle $R_j = (\alpha_j, \beta_j) \times (\gamma_j, \delta_j)$ be

$$\begin{aligned} S^j_{k,\ell}(x_1, x_2) &= B_j(x_1, x_2) \sin\left[\left(k + \frac{1}{2}\right)\pi \frac{x_1 - \alpha_j}{\beta_j - \alpha_j}\right] \\ &\quad \times \sin\left[\left(\ell + \frac{1}{2}\right)\pi \frac{x_2 - \gamma_j}{\delta_j - \gamma_j}\right], \end{aligned} \tag{4.4.14}$$

where $\sum_j B_j^2 = 1$ and B_j and $B_{j'}$ are mirror images of each other around an edge common to R_j and $R_{j'}$.

If the boundary Γ is a closed surface on which a global parametrization does not exist, then Γ is decomposed into patches of roughly the same size plus a number of aberrant regions. On the equal-sized patches, an orthogonal basis can be taken. For an example where Γ is a sphere, see [2].

We now discuss a similar analysis procedure for the Helmholtz operator on a surface S,

$$\bar{H}_\lambda(f) = \int_S \frac{e^{i\lambda|s-y|}}{|x-y|} f(y)d\sigma(y), \tag{4.4.15}$$

$d\sigma(y)$ being the surface measure on S. The sparsity of the matrix corresponding to the operator $\bar{H}_\lambda$ is due to oscillations, when $|s - y|$ is "large."

We separate our discussion into two cases. The first case is for nonneighboring patches on which $|x-y|$ is "large" so that $\bar{H}_\lambda$ is oscillatory. Similar to what we did for the oscillatory integral operator H_λ, we have roughly $N^2 \approx \lambda^2$ discretization points, and the window size for each patch is $\eta \times \eta$ with $\eta \approx \lambda^{-1/2}$ and $\eta^{-2} \approx N$. Hence, each patch has $\frac{1}{\eta} \times \frac{1}{\eta}$ points.

Let Q_j and $Q_{j'}$ be two non-neighboring patches at a distance roughly $r_{j,j'}\eta$ from each other. Let Q_j and $Q_{j'}$ be parameterized in terms of the parameters s and t respectively. As before, we need to estimate

$$\frac{1}{\eta^4} \int_{Q_j} \int_{Q_{j'}} e^{i\lambda|x(s)-x(t)|} b_j(s) b_{j'}(t) e^{\pm i(k\cdot s + k'\cdot t)/\eta} d\sigma(s) d\sigma(t).$$

We rescale the above integral into an integral over the unit square $V_2 = [0, 1]^2$ as follows.

$$\int_{V_2} \int_{V_2} e^{i\lambda|x(a_j+\eta u)-x(a_{j'}+\eta v)|} b(u)b(v) e^{\pm i(ku+k'v)} du dv. \tag{4.4.16}$$

Denote $\bar{\Phi} = N|x(a_j + \eta u) - x(a_{j'} + \eta v)| \pm k \cdot u \pm k' \cdot v$. We compute its partial derivatives by using the following.

$$\begin{aligned} x(a_j + \eta u) &= x(a_j) + u \cdot \eta(u_1 \cdot \mathbf{T}_1 + u_2 \cdot \mathbf{T}_2) + O(\eta^2 u^2), \\ \frac{\partial x}{\partial u_i} &= \eta \mathbf{T}_i + O(\eta^2 u), \\ \frac{1}{\sqrt{N}} \frac{\partial \bar{\Phi}}{\partial u_i} &= \sqrt{N} \eta \mathbf{T}_i \frac{(x(a_j+\eta u) - x(a_{j'}+\eta v))}{|x(a_j+\eta u) - x(a_{j'}+\eta v)|} \\ &\quad + O\left(\frac{1}{\sqrt{N}}\right) \pm \frac{k_i}{\sqrt{N}}, \end{aligned}$$

where $u = (u_1, u_2)$ and $k = (k_1, k_2)$. For different u_i, $i = 1, 2$, the first terms are the coordinates of points on an ellipse of length about $1/r_{j,j'}$. Using procedures similar to those for H_λ, we eventually obtain

$$N\delta^2 \sum_{j'} \frac{N}{r_{j,j'}^2} \le c\delta^2 N^2 \log N$$

as the number of matrix entries resulting from nonneighbor interactions of the basis functions defined on "equal-sized" rectangular patches that should be kept.

For the interactions of the irregular patches basis with the more regular basis functions, we just use the gradient of the phase in u or v (as we did for the corners on a curve), leading to a total matrix having $CN^2 \log N$ entries.

We now consider the effect of the singularity on the interactions of a window with its close neighbors, or with itself. We need to compute

$$\int_{V_2} \int_{V_2} \frac{e^{i\lambda|x(a+\eta u) - x(a+\eta v)|}}{|x(a + \eta u) - x(a + \eta v)|} b(u)b(v) e^{\pm i(k\cdot u + k'\cdot v)} du dv. \tag{4.4.17}$$

We can easily estimate (4.4.17) by parameterizing the surface in terms of its tangent plane at the center of the box. The parametrization gives

$$|x(a + \eta u) - x(a + \eta v)| = \eta|u - v| \left(1 + \eta\bar{\Phi}_\eta(u, v)\right),$$

where $\bar{\Phi}_\eta(u, v)$ has bounded derivatives.

We next compute

$$\sqrt{N} \int_{V_2} \int_{V_2} \frac{e^{i\sqrt{N}|u-v|}}{|u - v|} \beta_N(u, v) e^{\pm i(k\cdot u + k'\cdot v)} du dv, \tag{4.4.18}$$

where $\beta_N(u, v)$ has derivatives bounded independently of N and is supported in $|u| \le 1$, $|v| \le 1$.

The following lemma allows us to prove that the matrix is concentrated near the entries $k = \pm k'$.

Lemma 4.4.1 *Let $\zeta \in L^1(\mathbb{R}^2)$ and $\beta \in L^1(\mathbb{R}^4)$. Then the dimensionality-reducing formula*

$$\int_{\mathbb{R}^4} \zeta(u-v)\beta(u,v)e^{i(k\cdot u+k'\cdot v)}dudv = \int_{\mathbb{R}^2} \hat{\zeta}(\xi)\hat{\beta}(k-\xi, k'+\xi)d\xi,$$

holds. Here, $\hat{\zeta}$ and $\hat{\beta}$ are the Fourier transforms of ζ in $\mathbb{R}^2$ and β in $\mathbb{R}^4$ respectively.

Proof. The lemma can be proved by directly using the Plancherel theorem. □

In integral (4.4.18), we denote $\zeta_N(u-v) = \sqrt{N}e^{i\sqrt{N}|u-v|}\chi_2(u-v)/|u-v|$, where

$$\chi_2(t) = \begin{cases} 1, & \text{if } |t| < 2, \\ 0, & \text{if } |t| > 2. \end{cases}$$

Then $|\widehat{\beta_N}(k-\xi, k'+\xi)| \leq C_M/(1+|k-\xi|+|k'+\xi|)^{M+1}$ for each $M > 0$ and $|\widehat{\zeta_N}(\xi)| \leq C\sqrt{N}$. Hence, we can estimate integral (4.4.18) for the (k, k') entry as

$$\sqrt{N}\int_{\mathbb{R}^1} \frac{C_M}{(1+|k-\xi|+|k'+\xi|)^{M+1}}d\xi \leq \frac{C_M\sqrt{N}}{1+|k+k'|^M}.$$

In addition, we count the total number of terms in self window interactions to be N^2. Taken together, for the N^2 discretization points, we obtain a matrix having the order $CN^2\log N$.

4.5 DRE construction and numerical integration using measure theory

In this section, we will describe a method given by Burrows [6] for constructing DREs using *measure theory* and *Lebesgue integration.* The DRE reduces a multiple Lebesgue integral into a one-dimensional integral of a measure function. The method is particularly useful for integrands that are highly oscillatory in character or are singular, so that it offers another way to approximate oscillatory integrals, which we discussed in Chapter 3. By using *Burrows' DRE,* we can also derive a technique of numerical integration. We begin by looking at the one-dimensional integral of the form

$$\int_a^b f(x)dx, \tag{4.5.1}$$

where $f(x)$ is a bounded function on $[a, b]$ such that

$$0 \leq y_0 \leq f(x) \leq y_N. \tag{4.5.2}$$

The definition of the Lebesgue integral is expressed in terms of the measure of the set of points included in $I = [a, b]$. We define the measure of I to be $m(I) = b-a$, and for any open interval $I_1 = (\alpha, \beta) \subset I$ we define $m(I_1) = \beta-\alpha$. For a large set of points $E \subset I$ the exterior measure of E is defined as

$$m_e(E) = sup\ m(\cup_i J_i), \tag{4.5.3}$$

where $\{J_i\}$ is a set of open intervals such that $E \subset \cup_i J_i$.

By the Heine–Borel theorem we can choose a finite number of such open intervals. The set E is said to be measurable if

$$m_e(E) + m_e(\bar{E}) = b - a, \tag{4.5.4}$$

where $\bar{E} = I - E$, and the measure is then identical to the exterior measure.

For a nonnegative bounded measurable function $f(x)$, we denote the measure of the set of points in I for which $f(x) \geq y$ by $\mu(y)$. This is called a *measure function.* Obviously, this function is a monotonically decreasing function that satisfies

$$\mu(y_N) = 0 \leq \mu(y) \leq b - a = \mu(y_0). \tag{4.5.5}$$

Thus, the Lebesgue integral of $f(x)$ over I may be defined as

$$\int_a^b f(x)dx = y_0\mu(y_0) + \lim_{\substack{n\to\infty \\ \max \lambda_i \to 0}} \sum_{i=1}^{n} \mu(y_i)(y_i - y_{i-1}), \tag{4.5.6}$$

where $\lambda_i = y_i - y_{i-1}$. We can rewrite this as

$$\int_a^b f(x)dx = y_0\mu(y_0) + \int_{y_0}^{y_N} \mu(y)dy, \tag{4.5.7}$$

where the integral of $\mu(y)$ is defined in the sense of Riemann integration and exists because the limit in (4.5.6) exists (see Mikhlin [56] and Riesz and Nagy [59]).

Therefore, to estimate the integral $\int_a^b f(x)dx$, we need to find exact or approximate $\mu(y)$ for all y and compute the integral

$$\int_{y_0}^{y_N} \mu(y)dy \tag{4.5.8}$$

using the Gaussian or *Newton–Cotes formula.*

The restriction $f(x) \geq 0$ may be removed by defining

$$f_1(x) = \frac{1}{2}(|f(x)| + f(x)) \geq 0 \tag{4.5.9}$$

and

$$f_2(x) = \frac{1}{2}(|f(x)| - f(x)) \geq 0, \tag{4.5.10}$$

so that we may define

$$\int_a^b f(x)dx = \int_a^b f_1(x)dx - \int_a^b f_2(x)dx. \qquad (4.5.11)$$

Example 4.5.1 As an example, let us consider the integral $\int_1^5 \log x dx$. Since $\log x$ is monotonically increasing, we have $y_0 = 0$, $y_N = 1.609438$, and

$$\mu(y) = 5 - e^y, \quad y_0 \leq y \leq y_N.$$

Using the five-node Simpson's rule with $h = (y_N - y_0)/4 = 0.4023595$, we estimate $\int_0^{1.609438} \mu(y)dy$ to be 4.046622, while the actual value of $\int_1^5 \log x dx$ is 4.04719. In addition, if the five-node Simpson's rule is applied to $\int_1^5 \log x dx$ directly, then we have the less accurate estimate 4.0414773. For more examples, see [6].

It is often very hard or even impossible to find an explicit formula for $\mu(y)$. Another problem arises when y_0 and y_N are not known explicitly and have to be estimated, producing unknown errors in the numerical results. When $\mu(y)$ cannot be calculated exactly, we can divide I into m subsets of points. Each subset, I_i, has the measure p_i, where $\sum_{i=1}^m p_i = b - a$. Next we select $n \geq m$ points x_j $(j = 1, \dots, n)$ in I such that there is at least one point in each I_i. The function values, $f(x_j)$, at these n points are then calculated. Given any $\bar{y}$ with

$$y_0 \leq \bar{y} \leq y_N,$$

we approximate $\mu(\bar{y})$ by

$$\mu(\bar{y}) = \sum_{i=1}^m \epsilon_i p_i, \qquad (4.5.12)$$

where

$$\begin{aligned} \epsilon_i &= 0, && \text{if } \bar{y} > \max_j f(x_j), \quad x_j \in I_i, \\ \epsilon_i &= 1, && \text{if } \bar{y} < \min_j f(x_j), \quad x_j \in I_i, \\ 0 \leq \epsilon_i &\leq 1, && \text{otherwise.} \end{aligned} \qquad (4.5.13)$$

One possible way to choose ϵ_i when

$$\min_j f(x_j) \leq \bar{y} \leq \max_j f(x_j)$$

is the linear approximation

$$\epsilon_i = \frac{\max_j f(x_j) - \bar{y}}{\max_j f(x_j) - \min_j f(x_j)}, \quad x_j \in I_i.$$

In particular, if $n = m$, we take

$$\epsilon_i = \begin{cases} 1, & \text{if } f(x_i) \geq \bar{y},\ x_i \in I_i, \\ 0, & \text{otherwise.} \end{cases} \tag{4.5.14}$$

The approximation can be improved using the *equidistributed sequence* $\{x_i\}$. An example of such a sequence of points on [0, 1] is

$$x_i = \langle i\theta \rangle \equiv i\theta - [i\theta],$$

$\langle t \rangle$ being the fractional part of t (see [18]). For any bounded Riemann integrable function $g(x)$, $x \in [a, b]$, an equidistributed sequence $\{x_i\}$ in $[a, b]$ yields

$$\lim_{n\to\infty} \left(\frac{b-a}{n}\right)\left(\sum_{i=1}^{n} g(x_i)\right) = \int_a^b g(x)dx. \tag{4.5.15}$$

By choosing

$$g(x) = \begin{cases} 1, & \text{if } f(x) \geq \bar{y}, \\ 0, & \text{otherwise,} \end{cases} \tag{4.5.16}$$

we obtain

$$\lim_{n\to\infty} \frac{N_n}{n} = \frac{\mu(\bar{y})}{b-a}, \tag{4.5.17}$$

where N_n is the number of x_i $(i = 1, \ldots, n)$ for which $f(x_i) \geq \bar{y}$.

We now consider the multiple integral

$$\int_\Omega f(X)dX, \tag{4.5.18}$$

which is defined in the Lebesgue sense analogously to the one-dimensional case. We also define the measure of the finite region Ω by $m(\Omega) = V$, which is the volume of Ω. For instance,

$$S = \{X \in \Omega / |X - X_0| < r\} \subset \mathbb{R}^3$$

is an open sphere of measure $m(S) = 4\pi r^3/3$.

In parallel with the one-dimensional case, denote

$$\mu(y) = \{X | X \in \Omega, f(X) \geq y\}. \tag{4.5.19}$$

We have the exact DRE

$$\begin{aligned} \int_\Omega f(X)dX &= y_0\mu(y_0) + \lim_{\substack{n\to\infty \\ \max \lambda_i \to 0}} \sum_{i=1}^{n} \mu(y_i)(y_i - y_{i-1}) \\ &= y_0\mu(y_0) + \int_{y_0}^{y_N} \mu(y)dy. \end{aligned} \tag{4.5.20}$$

(4.5.18) is thus reduced to a one-dimensional integral $\int_{y_0}^{y_N} \mu(y)dy$.

In most cases, $\mu(y)$ cannot be calculated exactly; so an estimate is needed. Similar to the one-dimensional case, we can separate Ω into several subsets I_i. This treatment is advantageous if I_i are bounded by the contours of $f(X)$ (i.e., $f(X) = C$, C being a constant), since in these cases $\max_{X_i \in I_i} f(X_i)$ and $\min_{X_i \in I_i} f(X_i)$ are known exactly.

For convenience, we pick the points $X_i = (x_{1i}, x_{2i}, \dots, x_{si})$ from equidistributed sequences of points in $\Omega \in \mathbb{R}^s$. For instance, if $\Omega = [0, 1]^s$, then the sequence $X_i = (x_{1i}, x_{2i}, \dots, x_{si})$ with

$$x_{ji} = \langle i\theta_j \rangle \equiv i\theta_j - [i\theta_j] \tag{4.5.21}$$

($j = 1, \dots, s$) is an equidistributed sequence. Here θ_j are s independent irrational numbers, which implies that

$$\sum_{j=1}^{s} a_j \theta_j \neq 0 \tag{4.5.22}$$

for any set a_j of rational numbers (see [18] and [12]).

Example 4.5.2 We now consider the integral

$$\begin{aligned} J &= \int_{\Omega} f(X)dX \\ &= \frac{1}{\pi^3} \int_0^{\pi} \int_0^{\pi} \int_0^{\pi} \frac{dxdydz}{3.75 - \cos x - \cos y - \cos z} \\ &\approx 0.30781. \end{aligned} \tag{4.5.23}$$

The integrand is well represented by polynomials, and a conventional 27-point Gaussian rule (with three points in each dimension) produces the approximation value 0.30786. However, the convergence is slow as 729 points yield the value 0.30780 (see [18]).

We will apply Burrows' DRE to give a numerical integration. Taking approximations (4.5.7) and (4.5.14), choosing eight data points

$$X_i = \left(k\frac{\pi}{4}, j\frac{\pi}{4}, \ell\frac{\pi}{4}\right), \quad k, j, \ell = 1, 3,$$

(here I_i are eight cubes centered by X_i and are of measure $p_i = \pi/8$ for all i), and applying a five-point Gaussian quadrature formula for the corresponding integral $\int_{y_0}^{y_N} \mu(y)dy$, we obtain an approximation value, 0.3064529. Note that y_N and y_0 can be calculated exactly here because $f(X) = f(x, y, z)$ is monotonically decreasing and $f(0, 0, 0) \approx 1.\dot{3}$ and $f(\pi, \pi, \pi) \approx 0.148148$.

If we use the equidistributed sequence $\{X_i\}$ with

$$X_i = (x_i, y_i, z_i) = (i\sqrt{2}, i\sqrt{3}, i\sqrt{7}), \tag{4.5.24}$$

we obtain the estimates $J_n \approx J$ (cardinal number of data points $n = 30, 35, 40, 45$) shown in the following table.

number of data points n	J_n
30	0.3095236
35	0.3065273
40	0.3078608
45	0.3088979

Obviously, larger n produces significantly better estimates for this simple approximation, and we cannot hope to match the results obtained by more conventional methods when the integrands can be represented accurately by polynomials. Therefore, to improve the results, we need a better estimate of $\mu(y)$.

Let

$$L(f;k) = \mu(y_0) - k(f - y_0), \tag{4.5.25}$$

$k \geq 0$. This is a linear monotonically decreasing function of f with $L(y_0;k) = \mu(y_0)$.

Suppose

$$L(f;k) > \mu(f), \quad y_0 < f \leq y_N \tag{4.5.26}$$

and $\hat{f}$ satisfies

$$L(\hat{f};k) = 0. \tag{4.5.27}$$

Since $L(y_N;k) > \mu(y_N) = 0$ and $L(f;k)$ is monotonically decreasing, then $\hat{f} > y_N$ and

$$\int_{y_0}^{\hat{f}} L(f;k)df > \int_{y_0}^{y_N} \mu(y)dy. \tag{4.5.28}$$

Similarly, if

$$L(f;k) < \mu(f), \quad y_0 < f \leq y_N, \tag{4.5.29}$$

then

$$\int_{y_0}^{\hat{f}} L(f;k)df < \int_{y_0}^{y_N} \mu(y)dy. \tag{4.5.30}$$

Denote

$$I(k) = \int_{y_0}^{\hat{f}} L(f;k)df. \tag{4.5.31}$$

$I(k)$ is a continuous function of k $(0 \le k < \infty)$. Assuming there exist k_1 and k_2 so that $L(f; k_1)$ and $L(f; k_2)$ satisfy (4.5.26) and (4.5.28), respectively, then there exists k_0 such that

$$I(k_0) = \int_{y_0}^{y_N} L(f; k)df. \tag{4.5.32}$$

This assumption is valid for $y_N > y_0$ provided that $m(X|X \in \Omega, f(X) = y_0) = 0$. Furthermore,

$$L(f_1; k_0) = \mu(f_1) \tag{4.5.33}$$

holds for some f_1, $y_0 < f_1 \le y_N$. From this relation, we find $k_0 = (\mu(y_0) - \mu(f_1))/(f_1 - y_0)$. Thus, we can find f_2, $y_0 < f_2 \le y_N$, which yields an approximation of the form

$$I(k) \approx \int_{y_0}^{y_N} \mu(y)dy \tag{4.5.34}$$

for $\bar{k} = (\mu(y_0) - \mu(f_2))/(f_2 - y_0)$; i.e.,

$$L(f; \bar{k}) = \mu(f_2).$$

Obviously, if $f_2 = f_1$, then equation (4.5.34) is exact. The most economical choice of f_2 is the value for which

$$\mu(f_2) = \mu(y_0)/2. \tag{4.5.35}$$

In this case, $\bar{k} = \mu(y_0)/2(f_2 - y_0)$ and

$$L(f_2; \bar{k}) = \mu(y_0) - \frac{\mu(y_0)}{2(f_2 - y_0)}(f - y_0). \tag{4.5.36}$$

The root of $L(f; \bar{k}) = 0$ is then $\hat{f} = y_0 + 2(f_2 - y_0)$. Direct computing yields

$$\int_{y_0}^{\hat{f}} L(f; \bar{k})df = (f_2 - y_0)\mu(y_0). \tag{4.5.37}$$

Hence, from equations (4.5.31), (4.5.34), and (4.5.37) we obtain

$$\begin{aligned} \int_{\Omega} f(X)dX &= y_0\mu(y_0) + \int_{y_0}^{y_N} \mu(y)dy \\ &\approx f_2\mu(y_0). \end{aligned} \tag{4.5.38}$$

The above approximation (4.5.38) is the most economical because only one point, $(f_2, \mu(y_0)/2)$, is needed.

We can use the value, f_3, of $f(X)$ at the center of the cube to approximate f_2. Any plane passing through the center of the cube divides the volume of the cube into equal parts. Also, the plane tangent to the surface $f(X) = f_3$ goes through the center. If this tangent plane approximates the surface $f(X) = f_3$ accurately and $f(X)$ is monotonic, then $\mu(f_3) \approx \mu(y_0)/2$ and, consequently, $f_2 \approx f_3$. From approximation (4.5.38), we have $J \approx 0.2666\dot{7}$. This can be improved by dividing the cube into subcubes, so that the tangent plane approximates the surface more closely, and integrating over each subcube.

For convenience, we divide the cube into n^3 equal subcubes. Since $f(X)$ is symmetrical in x, y, and z, the number of function evaluations can be decreased. In the following table, we give the approximations $J_n \approx J$, the number n, and the number of function evaluations.

n	J_n	No. of function evaluations
2	0.3054125	4
4	0.307796	20
5	0.307806	35
6	0.307807	56
8	0.307808	120

As an example, $n = 5$ requires division into 125 subcubes, but the symmetry allows us to obtain the estimate by using only 35 evaluations. Even when the symmetry is ignored, 125 function evaluations still compares favorably to the 729 required for less accurate conventional methods.

4.6 Error analysis of numerical integration

In this section, we will discuss the error and convergence of the numerical integration that is from Burrows' exact DRE (4.5.20). Since the DRE is exact, the errors in evaluating the multiple integral $\int_\Omega f(X)dX$, other than those due to numerical roundoff, arise from the following sources:

(i) the error from the method used to estimate

$$\int_{y_0}^{y_n} \mu(y)dy; \tag{4.6.1}$$

(ii) the error from estimating $\mu(\bar{y})$ for some $\bar{y}$, when necessary;

(iii) the error from estimating y_0 and y_N when they are unknown.

To examine the errors in (i), we need to consider the form of $\mu(y)$. If $\mu(y)$ is sufficiently differentiable (see Example 4.5.1), we can use the Newton–Cotes or Gaussian rules to estimate integral (4.6.1). Analyzing the error terms that include derivatives of $\mu(y)$ shows that the approximation will converge as the number of points increases. However, $\mu(y)$ is often not differentiable. In cases where $\mu(y)$

is continuous but not necessarily differentiable, by Jackson's Theorem (see [58] and [75]) there exists a polynomial $P_n(y)$ for each $n = 1, 2, \dots$ such that

$$|\mu(y) - P_n(y)| < 6\omega\left(\frac{y_N - y_0}{2n}\right), \tag{4.6.2}$$

where $y_0 < y < y_N$, $n = 1, 2, 3, \dots$, and $\omega(\delta)$ is the modulus of continuity of $\mu(y)$; i.e.,

$$\omega(\delta) = \max_{|y_1 - y_2| < \delta} |\mu(y_1) - \mu(y_2)|.$$

Consequently, for any $\epsilon > 0$ and sufficiently large n, we have $|\mu(y) - P_n(y)| < \epsilon$.

Suppose now that integral (4.6.1) is approximated using the Newton–Cotes or Gaussian rule of the form

$$\sum_{i=1}^{m} \alpha_i \mu(y_i)$$

with degree of algebraic precision n. Denote

$$E = \int_{y_0}^{y_N} \mu(y)dy - \sum_{i=1}^{m} \alpha_i \mu(y_i). \tag{4.6.3}$$

Then from $\int_{y_0}^{y_N} P_n(y)dy = \sum_{i=1}^{m} \alpha_i P_n(y_i)$ we have

$$|E| < \int_{y_0}^{y_N} |\mu(y) - P_n(y)|dy + \sum_{i=1}^{m} |\alpha_i||\mu(y_i) - P_n(y_i)|. \tag{4.6.4}$$

It follows that

$$|E| < (y_N - y_0)\epsilon + \sum_{i=1}^{m} |\alpha_i|\epsilon.$$

If $\alpha_i > 0$ for all i, then

$$\sum_{i=1}^{m} |\alpha_i| = \sum_{i=1}^{m} \alpha_i = y_N - y_0.$$

Therefore, $|E| < 2(y_N - y_0)\epsilon$. Thus for sufficiently large n the above procedure converges.

If $\mu(y)$ is not continuous, we can use the technique shown in Example 4.5.2 to estimate integral (4.6.1). Its convergence will be established in the following theorem.

Theorem 4.6.1 *Let p be the measure of each subcube into which the cube with volume V has been divided. X_r is the rth subcube, and y_N^r and y_0^r denote the maximum and minimum values of y in the rth subcube. Then we have*

$$\left| \int_{y_0}^{y_N} \mu(y)dy - \sum_{r=1}^{n} f(X_r)p \right| < \theta \max_{1\le r\le n} (y_N^r - y_0^r)V \tag{4.6.5}$$

and

$$\mu(f(X_r)) = \frac{p}{2}, \tag{4.6.6}$$

where $0 < \theta < 1$.

Proof. The error in the approximation over the rth subcube is

$$\begin{aligned} E_r &= \int_{y_0^r}^{y_N^r} \mu(y)dy - \int_{y_0^r}^{\hat{f}^r} L(f;\bar{k}^r)df \\ &= \int_{y_0^r}^{\bar{f}^r} (\mu(f) - \hat{L}(f;\bar{k}^r))df, \end{aligned} \tag{4.6.7}$$

where $\bar{k}^r = \mu(y_0^r)/2(f_2^r - y_0^r)$ $(y_0^r \le f_2^r \le y_N^r)$, $\bar{f}^r = \max\{\hat{f}^r, y_N^r\}$, and

$$\hat{L}(f;\bar{k}^r) = \begin{cases} L(f;\bar{k}^r), & y_0^r \le f < \hat{f}^r, \\ 0, & \text{otherwise.} \end{cases}$$

Since $L(y_0^r;\bar{k}^r) = \mu(y_0^r)$ and $L(f_2^r;\bar{k}^r) = \mu(f_2^r) = p/2$,

$$|\mu(f) - \hat{L}(f;\bar{k}^r)| < p/2.$$

In addition, because $\hat{f}^r$ is the root of

$$L(f;\bar{k}^r) = \mu(y_0^r) - \bar{k}^r(f - y_0^r),$$

we have

$$\hat{f}^r = \frac{\mu(y_0^r)}{\bar{k}^r} + y_0^r = 2f_2^r - y_0^r.$$

Thus $\bar{y}^r = \max\{2f_2^r - y_0^r, y_N^r\}$ and

$$|E_r| < \frac{1}{2}(\bar{f}^r - y_0^r)p < \theta_r(y_N^r - y_0^r)p$$

for some θ_r, $0 < \theta_r < 1$.

Summing over all the subcubes,

$$\begin{aligned}\left|\int_{y_0}^{y_N} \mu(y)dy - \sum_{r=1}^{n} f(X_r)p\right| &= \left|\sum_{r=1}^{n} E_r\right| \\ &< \max_{1\le r\le n} \theta_r (y_N^r - y_0^r) \sum_{r=1}^{n} p \\ &< \theta \max_{1\le r\le n} (y_N^r - y_0^r) V \end{aligned} \tag{4.6.8}$$

for some θ, $0 < \theta < 1$. □

Remark 4.6.1. If $\mu(y)$ is continuous, then (4.6.5) shows the convergence because

$$\lim_{n\to\infty} \max_{1\le r\le n} (y_N^r - y_0^r) = 0. \tag{4.6.9}$$

(4.6.9) also holds when for a finite number of values, g_i, there exists a subregion of Ω for which $f(X) = g_i$. In such case $\mu(g_i)$ is discontinuous for all i and $\mu(y)$ is only sectionally continuous.

We now turn to the second source of error, from estimating $\mu(\bar{y})$. The approximations fall into three classes. If approximations (4.5.12) and (4.5.14) (or their higher dimensional analogues) are applied, in which the equidistributed points are taken, then the convergence is ensured. When approximations (4.5.12) and (4.5.13) are used, we can also obtain the convergence from the following theorem. As for the third class of approximations, the one shown in Example 4.5.2, its convergence will be discussed in Theorem 4.6.3.

Theorem 4.6.2 *Let $f(X)$ be bounded and Riemann integrable over Ω, which is of finite measure. If*

$$\Omega = \cup_{i=1}^{n} I_i, \quad m(I_i) = p_i, \tag{4.6.10}$$

then

$$\mu(\bar{y}) = \lim_{\substack{n\to\infty \\ \bar{p}\to 0}} \left(\sum_{i=1}^{n} \epsilon_i p_i\right), \tag{4.6.11}$$

where $\bar{p} = \max_{1\le i\le n} p_i$ and the ϵ_i are chosen as in (4.5.13).

Proof. Since $f(X)$ is Riemann integrable over Ω, the set B on which f is discontinuous is of measure zero. In the following discussion, we assume all elements of B have been removed from the various sets, leaving the measures of such sets unaltered.

As was described in (4.5.12) and (4.5.13), we have a set of data points $f(X_i)$ $(i = 1, \ldots, m)$, with at least one in each I_i.

Denote

$$\sup_{X \in I_i} f(X) = M_i \text{ and } \inf_{X \in I_i} f(X) = m_i. \tag{4.6.12}$$

Furthermore, define two step functions

$$\underline{f}(X) = \sum_{i=1}^{m} m_i \chi_{I_i}(X) \text{ and } \bar{f}(X) = \sum_{i=1}^{m} M_i \chi_{I_i}(X), \tag{4.6.13}$$

where

$$\chi_{I_i}(X) = \begin{cases} 1, & X \in I_i, \\ 0, & \text{otherwise.} \end{cases}$$

We also define two functions analogous to $\mu(y)$,

$$\mu_{\underline{f}}(y) = m(X|\underline{f}(X) \geq y) \text{ and } \mu_{\bar{f}}(y) = m(X|\bar{f}(X) \geq y). \tag{4.6.14}$$

Thus,

$$\mu_{\bar{f}}(y) \leq \mu(y) \leq \mu_{\underline{f}}(y).$$

Estimating $\mu(\bar{y})$ with (4.5.12) and (4.5.13) and noting that

$$m_i \leq \min_{X_j \in I_i} f(X_j) \text{ and } M_i \geq \max_{X_j \in I_i} f(X_j),$$

we have

$$\mu_{\bar{f}}(\bar{y}) \leq \sum_{i=1}^{n} \epsilon_i p_i \leq \mu_{\underline{f}}(\bar{y}). \tag{4.6.15}$$

When $n \to \infty$, $\max_{1 \leq i \leq n} p_i = \bar{p} \to 0$ and

$$\lim_{\substack{n \to \infty \\ \bar{p} \to 0}} \bar{f}(X) = \lim_{\substack{n \to \infty \\ \bar{p} \to 0}} \underline{f}(X) = f(X).$$

Hence,

$$\lim_{\substack{n \to \infty \\ \bar{p} \to 0}} \mu_{\bar{f}}(\bar{y}) = \lim_{\substack{n \to \infty \\ \bar{p} \to 0}} \mu_{\underline{f}}(\bar{y}) = \mu(\bar{y}). \tag{4.6.16}$$

It follows that

$$\lim_{\substack{n \to \infty \\ \bar{p} \to 0}} \left(\sum_{i=1}^{n} \epsilon_i p_i \right) = \mu(\bar{y}),$$

completing the proof. □

The third class of approximations was shown in Example 4.5.2. Similar to the approximation for f_3, we establish

$$f(X_c) \approx f(\hat{X}), \tag{4.6.17}$$

where X_c is the center of a subcube and $f(\hat{X})$ satisfies

$$\mu(f(\hat{X})) = p/2,$$

p being the measure of the subcube.

Assume $f(X)$ is differentiable. Then the equation of the tangent plane, T, of $f(X)$ at $X = X_c$ is

$$\nabla f(X_c) \cdot (X - X_c) = 0.$$

For any point on the tangent plane, we have

$$f(X) = f(X_c) + \nabla f(Y) \cdot (X - X_c)$$

for some Y on T. We are ready to give the following theorem.

Theorem 4.6.3 *There exists $\hat{X}$ on the tangent plane of $f(X)$ at point X_c such that*

$$\mu(f(\hat{X})) = \frac{p}{2}, \tag{4.6.18}$$

and

$$|f(\hat{X}) - f(X_c)| < kh, \tag{4.6.19}$$

where h is the length of each side of the cube and k is a fixed constant.

Proof. Let

$$\bar{f} = \sup_{X \in T} f(X) \text{ and } \underline{f} = \inf_{X \in T} f(X),$$

T denoting the plane through X_c tangent to the surface $f(X) = f(X_c)$. Since $f(X)$ is monotonic and the tangent plane separates the cube into two equal parts,

$$\mu(\bar{f}) \leq \frac{p}{2} \leq \mu(y).$$

Thus, there exists $\hat{X}$ such that

$$\mu(f(\hat{X})) = \frac{p}{2}.$$

$\hat{X}$ also satisfies $f(\hat{X}) = f(X_c) + \nabla f(Y) \cdot (\hat{X} - X_c)$ for some Y on T. It follows that

$$|f(\hat{X}) - f(X_c)| \le \sup_X |\nabla f(X)| \|\hat{X} - X_c\|,$$

where $\|\hat{X} - X_c\|$ is the ℓ^2 norm of $\hat{X} - X_c$. Therefore

$$\|\hat{X} - X_c\| < h\sqrt{M}/2,$$

where h is the length of each side of the cube and M is the dimension of the space.

Denote

$$k = \frac{\sqrt{M}}{2} \sup_X |\nabla f(X)|.$$

Then

$$|f(\hat{X}) - f(X_c)| < kh.$$

This completes the proof of the theorem. □

Theorem 4.6.3 implies that $E(h) = f(\hat{X}) - f(X_c)$ satisfies $\lim_{h \to 0} E(h) = 0$.

In Example 4.5.2, we established approximations (4.6.17) for each subcube. Thus,

$$\sum_{i=1}^{n} f(X_i)p \approx \sum_{i=1}^{n} f(\hat{X}_i)p, \tag{4.6.20}$$

where X_i is the center of the ith subcube and $\hat{X}_i$ satisfies $\mu(f(\hat{X}_i)) = p/2$. We now apply Theorem 4.6.3 to analyze the error, $E = \sum_{i=1}^{n} E_i(h)p$. Let the measure of the original cube be V. Then

$$|E| \le \max_{1 \le i \le n} |E_i(h)| V.$$

The inequality shows that approximation (4.6.20) converges when $n \to \infty$; i.e., $h \to 0$.

We now estimate the error from source (iii). Let

$$y_N = \sup_{X \in A/B} f(X) \text{ and } y_0 = \inf_{X \in A/B} f(X),$$

A is a set in the cube, and $B \subset A$ is the set of all discontinuous points of $f(X)$. We assume that $f(X)$ is Riemann integrable, which gives $mB = 0$. Denote by $\{X_j\}$ $(1 \le j \le m)$ a set of data points in A/B. We have already divided the cube into n subregions I_i $(i = 1, \ldots, n)$, in each of which there is at least one data point. We now establish the following theorem.

Theorem 4.6.4 *Let* $\bar{p} = \max_{1 \le i \le n} p_i$*, and let* p_i *be the measure of the subregion* I_i*. Then*

$$\lim_{\substack{n \to \infty \\ \bar{p} \to 0}} \max_{1 \le j \le m} f(X_j) = y_N.$$

Proof. Since $f(X)$ is continuous and bounded on A/B, there exists a point $\bar{X}$ such that $f(\bar{X}) = y_N$. Here $\bar{X} \in I_i$ for some i. For sufficiently small p_i,

$$f(\bar{X}) - f(X_j) < \epsilon$$

holds for $X_j \in I_i$ and arbitrary $\epsilon > 0$. Therefore,

$$f(\bar{X}) - \epsilon < f(X_j) < \max_{X_j} f(X_j).$$

It follows that

$$f(\bar{X}) - \max_{X_j} f(X_j) < \epsilon.$$

Because ϵ is arbitrary, the theorem is proved. □

The convergence theorem for the estimate of y_0 can be established similarly.

The method described in Sections 4.5 and 4.6 is especially useful for integrands that cannot be approximated accurately by a polynomial. The major disadvantage of the method is the additional errors resulting from estimating the measure function $\mu(y)$. The problem is partially eliminated when a large number of data points are presented. Theorems 4.6.1–4.6.4 ensure convergence of the approximations used. However, the convergence rates depend on the type of the integrand.

Chapter 5

DREs Over Complex Domains

In this chapter, we will introduce P.J. Davis's result regarding the construction of DREs over complex domains (see [15]). From *Green's theorem*, by using Schwarz functions, a double integral of an *analytic function* over a complex domain can be reduced to a *contour integral*. A subsequent application of the *radius theory* to the contour integral yields a DRE for the double integral. Since for any given *harmonic function* there always exists an analytic function the real part of which is the given harmonic function, a DRE for an analytic function leads to a DRE for the corresponding harmonic function. In Section 1, we will discuss the general method for constructing this type of DRE. Section 2 will give applications of the DRE in constructing quadrature formulas. In Section 3, we try to find (if possible) regions over which some given DREs and/or quadrature formulas of double integrals hold. Section 4 will discuss some additional topics on DREs over complex domains such as the construction of a Schwarz function regular in a slit region and applications of DREs in Fourier expansion.

5.1 DREs of the double integrals of analytic functions

We first clarify some notations and formulas. In the complex plane, denote $z = x + iy$ and $\bar{z} = x - iy$. Hence $x = (z + \bar{z})/2$ and $y = (z - \bar{z})/(2i)$, and $dz = dx + idy$ and $d\bar{z} = dx - idy = \overline{dz}$. Applying operators

$$\frac{\partial}{\partial z} = \frac{1}{2}\left(\frac{\partial}{\partial x} - i\frac{\partial}{\partial y}\right), \quad \frac{\partial}{\partial \bar{z}} = \frac{1}{2}\left(\frac{\partial}{\partial x} + i\frac{\partial}{\partial y}\right) \tag{5.1.1}$$

to function $f(z) = u(x, y) + iv(x, y)$ yields

$$\frac{\partial f}{\partial z} = \frac{1}{2}(u_x + iv_x - iu_y + v_y), \tag{5.1.2}$$

and

$$\frac{\partial f}{\partial \bar{z}} = \frac{1}{2}(u_x + iv_x + iu_y - v_y). \tag{5.1.3}$$

If f is analytic, then $u_x = v_y$, $u_y = -v_x$, and

$$\frac{\partial f}{\partial z} = u_x + iv_x = f'(z), \quad \frac{\partial f}{\partial \bar{z}} = 0. \tag{5.1.4}$$

Writing $\overline{f(z)} = u(x, y) - iv(x, y)$, we obtain similarly

$$\frac{\partial \overline{f(z)}}{\partial z} = 0, \quad \frac{\partial \overline{f(z)}}{\partial \bar{z}} = \overline{f'(z)}. \tag{5.1.5}$$

If B is a point set of the complex plane, then $\bar{B} = \{\bar{z} : z \in B\}$ is called the reflection of set B. If $f(z)$ is analytic in the region B, then function

$$\bar{f}(z) = \overline{f(\bar{z})} \tag{5.1.6}$$

defined on $\bar{B}$ is analytic. Note that

$$\overline{f(z)} = \bar{f}(\bar{z}). \tag{5.1.7}$$

Let ∂B be the boundary of region B traversed in the positive sense, and let f and g be analytic in B and continuously differentiable on ∂B. In the following, we write several alternate forms of Green's theorem.

$$\iint_B \overline{f'(z)} g(z) dx dy = \frac{1}{2i} \int_{\partial B} \overline{f(z)} g(z) dz, \tag{5.1.8}$$

$$\iint_B \overline{f(z)} g'(z) dx dy = -\frac{1}{2i} \int_{\partial B} \overline{f(z)} g(z) \overline{dz}, \tag{5.1.9}$$

and

$$\begin{aligned} \iint_B \overline{f'(z)} g'(z) dx dy &= \frac{1}{2i} \int_{\partial B} \overline{f(z)} g'(z) dz \\ &= -\frac{1}{2i} \int_{\partial B} \overline{f'(z)} g(z) \overline{dz}. \end{aligned} \tag{5.1.10}$$

In particular, with $f(z) \equiv z$,

$$\iint_B g(z) dx dy = \frac{1}{2i} \int_{\partial B} \bar{z} g(z) dz. \tag{5.1.11}$$

We now introduce the *Schwarz function* of a plane curve. Suppose equation

$$\phi(x, y) = 0 \tag{5.1.12}$$

defines a curve C in the complex plane. Equation (5.1.12) can be written equivalently as

$$\phi\left(\frac{z+\bar{z}}{2}, \frac{z-\bar{z}}{2i}\right) = 0. \tag{5.1.13}$$

The coordinates z and $\bar{z}$ are sometimes referred to as the minimal or conjugate coordinates. Suppose that equation (5.1.13) has the solution

$$\bar{z} = S(z). \tag{5.1.14}$$

If ϕ is analytic, then S is analytic except possibly in the vicinity of certain singular points. $S(z)$ is called the Schwarz function of the curve C (see [16] and [13]). In addition, if ϕ is an algebraic function of x and y, then $S(z)$ is an algebraic function: it is an analytic multivalued function of z. If C is a closed analytic curve, then $S(z)$ has a single-valued analytic branch in an annulus-like region that contains C in its interior.

The following properties of the Schwarz function, $\bar{z} = S(z)$, are to be understood as holding along curve C. For $z = re^{i\theta}$,

$$r^2 = z\bar{z} = |S(z)|^2, \quad \theta = \frac{i}{2}\log\left(\frac{\bar{z}}{z}\right) = \frac{i}{2}\log\left(\frac{S(z)}{z}\right). \tag{5.1.15}$$

We also have

$$S'(z) = \frac{d\bar{z}}{dz} = \frac{dx - idy}{dx + idy} = \frac{1 - iy'}{1 + iy'}, \tag{5.1.16}$$

$$y' = -i\frac{1 - S'(z)}{1 + S'(z)}, \tag{5.1.17}$$

and

$$\left|S'(z)\right| = 1. \tag{5.1.18}$$

Equation (5.1.16) can be proved easily from the definition of the Schwarz function, and equations (5.1.17) and (5.1.18) are from (5.1.16) directly.

If a curve with the Schwarz function $S(z)$ passes through point z_0, then the tangent line of the curve at z_0 has equation

$$\bar{z} = S'(z_0)(z - z_0) + \bar{z}_0. \tag{5.1.19}$$

Let S and T be Schwarz functions of two curves with a common point z_1. Then the angle of intersection θ between the two curves at z_1 is

$$\tan\theta = i\left(\frac{S'(z_1) - T'(z_1)}{S'(z_1) + T'(z_1)}\right). \tag{5.1.20}$$

In particular, if $S'(z_1) = T'(z_1)$, then the two curves are tangent to each other at z_1, while if $S'(z_1) = -T'(z_1)$, the curves are orthogonal at z_1.

By taking derivatives of (5.1.17), we have

$$\frac{d^2y}{dx^2} = \frac{4iS''(z)}{(1+S'(z))^3} \tag{5.1.21}$$

along curve C. On the other hand, the arc measure ds is

$$\begin{aligned} ds &= \sqrt{dx^2+dy^2} = \sqrt{(dx+idy)(dx-idy)} \\ &= \sqrt{dzd\bar{z}} = \sqrt{S'(z)}dz. \end{aligned} \tag{5.1.22}$$

Hence, we obtain the curvature of C, k, as

$$k = iS''/2(S')^{3/2}. \tag{5.1.23}$$

From (5.1.18)

$$|k| = |S''|/2. \tag{5.1.24}$$

Therefore, we also call $S''(z)$ the complex curvature of C.

We now express the Schwarz function $S(z)$ of a closed analytic curve C in terms of the mapping function of C. Let $w = M(z)$ be an analytic function that performs a 1-1 conformal map of C and its interior onto the closed unit disc. Denote the inverse of $w = M(z)$ by $z = m(w)$. Then $S(z)$ can be written as

$$S(z) = \bar{m}(1/M(z)). \tag{5.1.25}$$

The point $z_R = \overline{S(z)}$ is the *Schwarzian reflection* of the point z in the analytic curve C. In view of this fact, $S(z)$ satisfies the functional equation

$$\bar{S}(S(z)) \equiv z. \tag{5.1.26}$$

We now discuss the *DREs of double integrals over complex regions.* For an analytic function $f(z)$, applications of formula (5.1.8) and the Schwarz function of ∂B yield the following DREs.

$$\begin{aligned} \iint_B \bar{z}^n f(z)dxdy &= \frac{1}{2i(n+1)}\int_{\partial B} \bar{z}^{n+1} f(z)dz \\ &= \frac{1}{2i(n+1)}\int_{\partial B} (S(z))^{n+1} f(z)dz, \end{aligned} \tag{5.1.27}$$

where $n \geq 0$ is an integer. We are now at liberty to change the path of integration, by Cauchy's theorem, or to apply *residue* calculus, so that we can change the last integral of equation (5.1.27) to a quadrature sum and/or a sum of definite integrals, which leads us to an exact DRE. Based on this idea, we will construct exact DREs over some special complex regions. In addition, separating the real part and imaginary part of the double integral over a complex region, we may obtain a pair of exact DREs for real functions.

The method for constructing DREs of double integrals of analytic functions can be modified for the double integral of a harmonic function (or the harmonic counterpart of a function). In fact, let $u(x, y)$ be harmonic in Q; we construct

$$v(x, y) = \int_{(x_0, y_0)}^{(x, y)} -u_y dx + u_x dy \tag{5.1.28}$$

for an arbitrary point (x_0, y_0) in Q. Then, $f(z) = u(x, y) + iv(x, y)$ is single-valued and analytic in Q. By using formula (5.1.27), one may obtain a DRE for the double integral $\int\int_Q f(z)dxdy = \int\int_Q [u(x, y) + iv(x, y)]dxdy$ and separate the real part of the DRE. This leads to an exact DRE for $\int\int_Q u(x, y)dxdy$.

We now give two applications of the above method for double integrals over some special regions. More applications will be shown in the next section. In all of these examples, $f(z)$ is always assumed to be analytic.

Example 5.1.1 For disc C: $|z - z_0| \le r$, all points on its boundary ∂C satisfy

$$(z - z_0)(\bar{z} - \bar{z}_0) = r^2.$$

Solving the above equation, we obtain the Schwarz function of ∂C:

$$\bar{z} = S(z) = \bar{z}_0 + r^2/(z - z_0), \tag{5.1.29}$$

where z_0 is a simple pole of $S(z)$. It can be proved that the circle is the only curve whose Schwarz function is a rational function of z. If $z_0 = 0$, $S(z) = r^2/z$. Substituting $S(z) = r^2/z$ into DRE (5.1.27), we obtain

$$\begin{aligned} \int\int_C \bar{z}^n f(z)dxdy &= \frac{\pi r^{2n+2} n!}{2\pi i(n+1)!} \int_{\partial C} \frac{f(z)}{z^{n+1}} dz \\ &= \frac{\pi r^{2n+2}}{(n+1)!} f^{(n)}(0) \end{aligned} \tag{5.1.30}$$

for $n = 0, 1, \ldots$ In particular, the case of n_0 gives the mean value theorem for analytic functions.

We now consider the half disc: $x^2 + y^2 \le r^2$, $y \ge 0$. Let ∂HC be the boundary of HC and R its circular arc traversed positively. From formula (5.1.27) we have

$$\begin{aligned} \int\int_{HC} f(z)dxdy &= \int_{\partial HC} \frac{1}{2i} \bar{z} f(z)dz \\ &= \frac{1}{2i} \int_{-r}^{r} x f(x)dx + \frac{1}{2i} \int_R \bar{z} f(z)dz. \end{aligned} \tag{5.1.31}$$

Substituting the Schwarz function $\bar{z} = S(z) = r^2/z$ into the last integral of the above equation and using the *Cauchy integral theorem,* we obtain

$$\begin{aligned}\frac{1}{2i}\int_R \bar{z} f(z)dz &= \frac{r^2}{2i}\int_R \frac{f(z)}{z}dz\\ &= \frac{r^2}{2i}\int_r^{\epsilon}\frac{f(x)}{x}dx + \frac{r^2}{2i}\int_{-\epsilon}^{-r}\frac{f(x)}{x}dx\\ &\quad+\frac{r^2}{2i}\int_{R'}\frac{f(z)}{z}dz,\end{aligned}\tag{5.1.32}$$

where R' designates a half circle of radius ϵ traversed positively. Notice

$$\frac{r^2}{2i}\int_{R'}\frac{f(z)}{z}dz = \frac{r^2}{2}\int_0^{\pi} f(\epsilon e^{i\theta})d\theta.$$

Therefore,

$$\lim_{\epsilon\to 0}\frac{r^2}{2i}\int_{R'}\frac{f(z)}{z}dz = \frac{\pi r^2}{2} f(0).\tag{5.1.33}$$

Finally, we combine equations (5.1.31), (5.1.32), and (5.1.33) and obtain the DRE

$$\begin{aligned}\iint_{HC} f(z)dxdy &= \frac{1}{2i}\int_{-r}^{r} x f(x)dx + \frac{r^2 i}{2}\int_{-r}^{r}\frac{f(x)}{x}dx\\ &\quad+\frac{\pi r^2}{2} f(0),\end{aligned}\tag{5.1.34}$$

where the second integral on the right-hand side of the above equation means the *Cauchy principal value.*

Example 5.1.2 For the ellipse E: $\frac{x^2}{a^2}+\frac{y^2}{b^2}\le 1$, $a > b$, the corresponding Schwarz function of its boundary curve ∂E is

$$S(z) = \frac{a^2+b^2}{a^2-b^2} + \frac{2ab}{b^2-a^2}\sqrt{z^2+b^2-a^2}.\tag{5.1.35}$$

Here, $z = \pm\sqrt{a^2-b^2}$ are branch points of $S(z)$. In the z plane cut from $-\sqrt{a^2-b^2}$ to $\sqrt{a^2-b^2}$, $S(z)$ can be defined as a single-valued analytic function.

Since the first term of $S(z)$ is a regular function in E, an application of DRE formula (5.1.27) gives

$$\begin{aligned}\iint_E f(z)dxdy &= \frac{ab}{i(b^2-a^2)}\int_{\partial E}\sqrt{z^2+b^2-a^2}\, f(z)dz\\ &= \frac{ab}{b^2-a^2}\int_{\partial E}\sqrt{a^2-b^2-z^2}\, f(z)dz.\end{aligned}\tag{5.1.36}$$

$\sqrt{a^2-b^2-z^2}$ is single-valued in the plane slit along $-\sqrt{a^-b^2} \leq x \leq \sqrt{a^2-b^2}$. Therefore, one may shrink the curve ∂E until it coincides with the slit traversed twice; the lower edge from $-\sqrt{a^2-b^2}$ to $\sqrt{a^2-b^2}$ and the upper edge back. On the first leg, $dz = dx$ and the radical is $-\sqrt{a^2-b^2-x^2}$; on the return, $dz = -dx$ and the radical is $\sqrt{a^2-b^2-x^2}$. This leads to the DRE formula

$$\iint_E f(z)dxdy$$
$$= \frac{2ab}{a^2-b^2} \int_{-\sqrt{a^2-b^2}}^{\sqrt{a^2-b^2}} \sqrt{a^2-b^2-x^2} f(x)dx. \tag{5.1.37}$$

If we choose $a = (\rho + \rho^{-1})/2$ and $b = (\rho - rho^{-1})/2$, then the corresponding $E \equiv E_\rho$ is the ellipse with foci at ± 1 and the semi-axis sum $a+b = \rho$. The DRE formula thus can be written accordingly as

$$\iint_{E_\rho} f(z)dxdy$$
$$= \frac{1}{2}(\rho^2 - \rho^{-2}) \int_{-1}^{1} \sqrt{1-x^2} f(x)dx. \tag{5.1.38}$$

Formula (5.1.38) was first obtained in [14] by using the double orthogonality of the Chebyshev polynomials of the second kind.

5.2 Construction of quadrature formulas using DREs

In this section, we will apply DRE formula (5.1.27) to construct quadrature formulas over some regions. Throughout, $f(z)$ is still assumed to be always analytic.

Example 5.2.1 The first region considered is bounded by the *rose curve* R_{2m}: $|z|^{2m} = c + b\cos 2m\theta$, $0 < |b| < c$, $m = 1, 2, \ldots$. Noting that

$$\cos n\theta = \frac{z^n + \bar{z}^n}{2(z\bar{z})^{n/2}},$$

we have

$$z^m \bar{z}^m = c + b\left(\frac{z^{2m} + \bar{z}^{2m}}{2z^m \bar{z}^m}\right).$$

Hence, the Schwarz function of R_{2m} is

$$\bar{z} = S(z) = z\left(\frac{c \pm \sqrt{c^2 - b^2 + 2bz^{2m}}}{2z^{2m} - b}\right)^{1/m}. \tag{5.2.1}$$

The case of $m = 1$, $c = a^2 + 2\epsilon^2$, and $b = 2\epsilon^2$ yields the bicircular quartic ∂Q:

$$|z|^2 \equiv r^2 = (a^2 + 2\epsilon^2) + 2\epsilon^2 \cos 2\theta = a^2 + 4\epsilon^2 \cos^2 \theta.$$

From (5.2.1), the corresponding Schwarz function of ∂Q is

$$S(z) = \frac{z(a^2 + 2\epsilon^2) + z\sqrt{a^4 + 4a^2\epsilon^2 + 4\epsilon^2 z^2}}{2(z^2 - \epsilon^2)}. \tag{5.2.2}$$

The quantity under the radical vanishes when $z = \pm i\sqrt{a^2 + (a^4/4\epsilon^2)}$. Since $a^4/4\epsilon^2 > 0$ and $r = a$ when $\theta = \pi/2, 3\pi/2$, these points lie outside the curve ∂Q. Hence, inside ∂Q, $S(z)$ can be defined as a single-valued analytic function with simple poles at $z = \pm\epsilon$.

We now denote the union of ∂Q and its interior by Q: $r^2 \leq a^2 + 4\epsilon^2 \cos^2 \theta$. From DRE formula (5.1.27), for $n = 0$ we obtain

$$\iint_Q f(z)dxdy = \frac{\pi}{2\pi i} \int_{\partial Q} S(z) f(z) dz, \tag{5.2.3}$$

where $S(z)$ is given by (5.2.2). The residue of $S(z)$ at the pole $z = \epsilon$ is

$$\begin{aligned} (z - \epsilon)S(z)|_{z=\epsilon} &= \frac{\epsilon(a^2+2\epsilon^2)+\epsilon\sqrt{a^4+4a^2\epsilon^2+4\epsilon^4}}{2(2\epsilon)} \\ &= \frac{a^2}{2} + \epsilon^2. \end{aligned}$$

A similar residue of $S(z)$ holds for pole $z = -\epsilon$. The residue theory yields the following quadrature formula.

$$\iint_Q f(z)dxdy = \pi \left(\frac{a^2}{2} + \epsilon^2 \right) (f(\epsilon) + f(-\epsilon)). \tag{5.2.4}$$

For $f(z) = z^{2\ell}$, we evaluate in polar coordinates the left-hand integral of (5.2.4),

$$\int_0^{2\pi} e^{i2\ell\theta} (a^2 + 4\epsilon^2 \cos^2 \theta)^{\ell+1} d\theta = 2\pi(\ell + 1)(a^2 + 2\epsilon^2)\epsilon^{2\ell}. \tag{5.2.5}$$

In particular, for $\ell = 0$, quadrature formula (5.2.4) gives the area of Q:

$$\iint_Q dxdy = \pi(a^2 + 2\epsilon^2).$$

We now construct the quadrature formula of integral $\iint_Q \bar{z}^n f(z)dxdy$ for all $n \geq 0$. Denote

$$N(z) = \frac{z}{2} \left((a^2 + 2\epsilon^2) + \sqrt{a^4 + 4a^2\epsilon^2 + 4\epsilon^2 z^2} \right). \tag{5.2.6}$$

Then $S^{n+1}(z) = N^{n+1}(z)/(z^2-\epsilon^2)^{n+1}$, which is regular inside ∂Q except at points $z=\pm\epsilon$, poles of S^{n+1} with order $n+1$. From DRE formula (5.1.27),

$$\begin{aligned}
&\iint_Q \bar{z}^n f(z)dxdy \\
= &\ \frac{\pi}{(n+1)!}\frac{n!}{2\pi i}\int_{\partial Q}\frac{N^{n+!}(z)}{(z-\epsilon)^{n+1}(z+\epsilon)^{n+1}}f(z)dz \\
= &\ \frac{\pi}{(n+1)!}\left\{\frac{d^n}{dz^n}\left(\frac{N^{n+1}(z)f(z)}{(z+\epsilon)^{n+1}}\right)|_{z=\epsilon}+\frac{d^n}{dz^n}\left(\frac{N^{n+1}(z)f(z)}{(z-\epsilon)^{n+1}}\right)|_{z=-\epsilon}\right\} \\
= &\ \frac{\pi}{(n+1)!}\Big[a_{n,0}f(\epsilon)+a_{n,1}f'(\epsilon)+\cdots+a_{n,n}f^{(n)}(\epsilon) \\
&+b_{n,0}f(-\epsilon)+b_{n,1}f'(-\epsilon)+\cdots+b_{n,n}f^{(n)}(-\epsilon)\Big],
\end{aligned} \tag{5.2.7}$$

where the constants $a_{n,k}$ and $b_{n,k}$ are independent of f and can be obtained explicitly by expanding the above bracket.

Let $u(x,y)$ be harmonic in Q. Then, as we described before, $f(z)=u(x,y)+iv(x,y)$ is analytic in Q if $v(x,y)=\int_{(x_0,y_0)}^{(x,y)}-u_ydx+u_xdy$. An application of formula (5.2.4) to $f(z)$ yields

$$\begin{aligned}
&\iint_Q(u+iv)dxdy \\
= &\ \pi\left(\frac{a^2}{2}+\epsilon^2\right)[u(\epsilon,0)+iv(\epsilon,0)+u(-\epsilon,0)+iv(-\epsilon,0)].
\end{aligned}$$

The real portion of the above equation is

$$\iint_Q u(x,y)dxdy = \frac{\pi}{2}\left(a^2+2\epsilon^2\right)(u(\epsilon,0)+u(-\epsilon,0)). \tag{5.2.8}$$

Equation (5.2.8) is an extension of the mean value theorem for harmonic functions. This identity leads to an inequality for harmonic functions. Let C_r be the circle $|z|\le r$. In view of the obvious inequalities $a^2+4\epsilon^2\ge a^2+4\epsilon^2\cos^2\theta\ge a^2$, we have the relation $C_a\subset Q\subset C_b$ for $b=\sqrt{a^2+4\epsilon^2}$. Thus, for the harmonic and nonnegative function u,

$$\iint_{C_b} udxdy \ge \iint_Q udxdy \ge \iint_{C_a} udxdy. \tag{5.2.9}$$

It is well known that a necessary and sufficient condition for u to be harmonic in a region G is that

$$u(z_0) = \frac{1}{\pi r^2}\iint_{|z-z_0|\le r} u(x,y)dxdy$$

for all $z_0\in G$ and all sufficiently small r. By using inequalities (5.2.9) and the above equation, we have

$$\pi(a^2+4\epsilon^2)u(0,0) \geq \frac{\pi}{2}(a^2+2\epsilon^2)\left(u(\epsilon,0)+u(-\epsilon,0)\right) \geq \pi a^2 u(0,0).$$

It follows that

$$\frac{a^2+4\epsilon^2}{a^2+2\epsilon^2}u(0,0) \geq \frac{1}{2}\left(u(\epsilon,0)+u(-\epsilon,0)\right) \geq \frac{a^2}{a^2+2\epsilon^2}u(0,0).$$

As another example of rose curves, we consider R_4: $r^4 \leq a^4 + 2b^4\cos 4\theta$, $a^4 > 2b^4$. The corresponding Schwarz function is given by

$$\bar{z}^2 = S^2(z) = \frac{a^4z^2 + z^2\sqrt{4b^4z^4+a^8-4b^8}}{2(z^4-b^4)}.$$

Inside R_4, $S^2(z)$ has only simple poles at points $z = \pm b, \pm bi$. The residue of S^2 at $z = b$ is

$$\frac{a^4z^2 + z^2\sqrt{4b^4z^4+a^8-4b^8}}{2(z+b)(z^2+b^2)} f(z)\mid_{z=b} = \frac{a^4}{4b}f(b).$$

The residues for the other poles are similar. Hence, from DRE (5.1.27), we find

$$\begin{aligned}\iint_{R_4} \bar{z}f(z)dxdy &= \tfrac{\pi}{2}\tfrac{1}{2\pi i}\int_{\partial R_4} S^2(z)f(z)dz \\ &= \tfrac{a^4\pi}{8b}[f(b) - if(ib) - f(-b) + if(-ib)].\end{aligned}$$

In view of the singularity of $S(z)$, one can construct a DRE of the integral $\int\int_{R_4} f(z)dxdy$ so that the integral is reduced to two line integrals from b to bi and from $-bi$ to $-b$. Thus, DREs of $\int\int_{R_4} f(z)dxdy$ have the following forms: for even values of n, $\int\int_{R_4} f(z)dxdy$ becomes a functional of that type while for odd n, it can be reduced to an interpolation functional.

Example 5.2.2 The equation of the *lemniscate* ∂U_r is

$$|p_n(z)| = r^n, \tag{5.2.10}$$

where $p_n(z) = (z-z_1)\cdots(z-z_n)$. On the geometry of these curves, see Walsh [68]. We rewrite equation (5.2.10) as

$$|p_n(z)|^2 = p_n(z)\overline{p_n(z)} = p_n(z)\overline{p_n}(\bar{z}) = r^{2n}. \tag{5.2.11}$$

If $p_n(z) = \sum_{j=0}^{n} a_jz^j$, then $\overline{p_n}(z) = \sum_{j=0}^{n} \overline{a_j}z^j$. Hence, on ∂U_r, we have

$$\overline{p_n}\,(S(z)) = r^{2n}/p_n(z). \tag{5.2.12}$$

The function $\overline{p_n}\,(S(z))$ has poles at $z_1, \ldots, z_n$, and $S(z)$ of ∂U_r is the algebraic function of z obtained by solving the above equation.

For instance, if $p_n(z) = z^n - 1$, then $\overline{p_n}(z) = z^n - 1$ and

$$\bar{z} = S(z) = \sqrt[n]{\frac{z^n + r^{2n} - 1}{z^n - 1}}. \tag{5.2.13}$$

For $n = 2$, equation (5.2.13) presents the *Cassinian oval*, ∂OC. If $r > 1$, the interior of ∂OC is denoted by OC. The Schwarz function of ∂OC,

$$\bar{z} = S(z) = \sqrt{\frac{z^2 + r^4 - 1}{z^2 - 1}},$$

has zeros at $\pm i\sqrt{r^4 - 1}$, which are exterior to OC. Therefore, $S(z)$ is single valued in the interior minus the cut $-1 \leq x \leq 1$. From formula (5.1.27) and the Cauchy integral theorem, we obtain

$$\begin{aligned} \iint_{OC} f(z)dxdy &= \frac{1}{2i}\int_{\partial OC}\sqrt{\frac{z^2 + r^4 - 1}{z^2 - 1}}\,f(z)dz \\ &= \int_{-1}^{1}\sqrt{\frac{x^2 + r^4 - 1}{1 - x^2}}\,f(x)dx. \end{aligned} \tag{5.2.14}$$

If $r < 1$, the *Cassinian oval* consists of two lobes. On the right lobe, $S(z)$ will have branch points at $z = 1$ and at $z = \sqrt{1 - r^4}$. Hence, we can obtain a DRE analogous to (5.2.14) over the right lobe that extends the real integral over $\sqrt{1 - r^4} \leq x \leq 1$. A similar DRE holds for the left lobe.

For the general lemniscate shown by (5.2.10), we first select a large enough r such that the locus consists of a closed curve containing $z_1, \ldots, z_n$ in its interior. Denote the region $|p_n(z)| \leq r^n$ by U_r. Since $\overline{p_n(z)} = r^{2n}/p_n(z)$ for all $z \in \partial U_r$, from (5.1.8) we have

$$\begin{aligned} \iint_{U_r} \overline{p_n'(z)}\,f(z)dxdy &= \frac{1}{2i}\int_{\partial U_r} \overline{p_n(z)}\,f(z)dz \\ &= \frac{r^{2n}\pi}{2\pi i}\int_{\partial U_r} \frac{f(z)}{p_n(z)}dz. \end{aligned} \tag{5.2.15}$$

If the z_i are distinct, from the residue theorem we obtain

$$\frac{1}{2\pi i}\int_{\partial U_r}\frac{f(z)}{p_n(z)}dz = \frac{f(z_1)}{p_n'(z_1)} + \cdots + \frac{f(z_n)}{p_n'(z_n)} = [f(z_1), \ldots, f(z_n)],$$

which is the nth divided difference of $f(z)$ with respect to the points $z_1, \ldots, z_n$. When we have multiple points, the contour integral equals the generalized divided difference. Therefore,

$$\iint_{U_r}\overline{p_n'(z)}f(z)dxdy = \pi r^{2n}[f(z_1), \ldots, f(z_n)]. \tag{5.2.16}$$

If the points $z_1, \cdots, z_n$ are uniformly spaced, then the divided difference becomes the ordinary nth order difference. For example, if $z_j = j-1$, $j = 1, \ldots, n$, then

$$\iint_{U_r}\overline{p_n'(z)}f(z)dxdy = \pi r^{2n}[f(0), \ldots, f(n-1)] = \pi r^{2n}\frac{\Delta^{n-1}f(0)}{(n-1)!}.$$

For another special case, if $f(z) = p_n'(z)$ in (5.2.15), we have

$$\iint_{U_r}|p_n'(z)|^2 dxdy = n\pi r^{2n}. \tag{5.2.17}$$

This equality gives the norm of the divided difference operator $D(f) := [f(z_1), \ldots, f(z_n)]$ for fixed points z_j, $j = 1, \ldots, n$. Hence,

$$D(f) = \frac{1}{\pi r^{2n}}\iint_{U_r}\overline{p_n'(z)}f(z)dxdy.$$

From [12], p. 218, the function $p_n'(z)/(\pi r^{2n})$ is the representer of the functional D over the Hilbert space $L^2(U_r)$, the space of all single-valued analytic functions f with $\int\int_{U_r}|f(z)|^2dxdy < \infty$. Therefore, the norm of D is given by

$$\begin{aligned}\| D \|_{U_r}^2 &= \iint_{U_r}\left|\frac{p_n'(z)}{\pi r^{2n}}\right|^2 dxdy \\ &= \frac{1}{\pi^2 r^{4n}}\iint_{U_r}|p_n'(z)|dxdy = \frac{n}{\pi r^{2n}}.\end{aligned}$$

It follows that $\| D \|_{U_r} = \sqrt{n/\pi}/r^n$.

If $p_N(z) = \Pi_{j=1}^n(z - z_j)^{\alpha_j}$, $j = 1, \ldots, n$, where $N = \sum_{j=1}^n \alpha_j$. Denote $|p_N(z)| \le r^N$ by $\tilde{U}_r$. Then similar to (5.2.15) and (5.2.16), we have

$$\begin{aligned}\iint_{\bar{U}_r} |p'_N(z)|^2 f(z)dxdy &= \frac{1}{2i}\int_{\partial\bar{U}_r} \overline{p_N(z)} p'_N(z) f(z)dz \\ &= \frac{\pi r^{2N}}{2\pi i}\int_{\partial\bar{U}_r} \frac{p'_N(z)}{p_N(z)} f(z)dz \\ &= \pi r^{2N}\sum_{j=1}^{n}\alpha_j f(z_j). \end{aligned} \tag{5.2.18}$$

At the end of this section, we use quadrature formula (5.2.16) to give a representation of interpolation remainders. Let $z_0, z_1, \ldots, z_n$ be points in the complex plane that are not necessarily distinct. Also, let $p_n(f; z)$ be the unique polynomial of degree no more than n that interpolates to an analytic function f at z_j, $j = 0, \ldots, n$. It is well known that the remainder of the interpolation $R_n(f; z) = f(z) - p_n(f; z)$ can be written as

$$R_n(f; z) = (z - z_0)\cdots(z - z_n)\,[f(z), f(z_0), \ldots, f(z_n)]\,. \tag{5.2.19}$$

Let z be an arbitrarily fixed point and $P(t) = (t - z)(t - z_0)\cdots(t - z_n)$, $t = u + iv$. Select a large enough r such that the set $\hat{U}_r$: $|P(t)| \le r^{n+2}$ contains $z, z_0, \ldots, z_n$ in its interior. From quadrature formula (5.2.16),

$$R_n(f; z) = \frac{(z - z_0)\cdots(z - z_n)}{\pi r^{2n+4}}\iint_{\hat{U}_r} \overline{P'(t)} f(t)dudv. \tag{5.2.20}$$

Then,

$$h(t) = \overline{\frac{(z - z_0)\cdots(z - z_n)}{\pi r^{2n+4}} P'(t)}$$

is the representer of the remainder functional $R_n(f; z)$ over the space $L^2(\hat{U}_r)$. Thus, from (5.2.17) we obtain

$$\begin{aligned}\| R_n \|^2_{\hat{U}_r} &= \iint_{\hat{U}_r} |h(t)|^2 dudv \\ &= \frac{|(z - z_0)\cdots(z - z_n)|^2}{\pi^2 r^{4n+8}}\iint_{\hat{U}_r} |P'(t)|^2 dudv \\ &= \frac{|(z - z_0)\cdots(z - z_n)|^2}{\pi^2 r^{4n+8}}(n + 2)\pi r^{2n+4}. \end{aligned}$$

It follows that

$$\| R_n \|_{\hat{U}_r} = \sqrt{\frac{n+2}{\pi}}\frac{|z - z_0)\cdots(z - z_n)|}{r^{n+2}}. \tag{5.2.21}$$

We now use DREs to construct quadrature formulas of double integrals over a region in the z plane, onto which the unit disc $|w| \leq 1$ can be 1-1 conformally mapped.

Let B be a simply connected region of the z plane containing $z = 0$, and suppose that $z = m(w)$, $m(0) = 0$, maps the unit circle $|w| \leq 1$ one-to-one conformally onto B. From DRE (5.1.27),

$$\begin{aligned}
&\iint_B \bar{z}^p f(z)dxdy \\
&= \frac{1}{2i(p+1)} \int_{\partial B} \bar{z}^{p+1} f(z)dz \\
&= \frac{1}{2i(p+1)} \int_{|w|=1} \overline{(m(w))}^{p+1} f(m(w))m'(w)dw \\
&= \frac{\pi}{p+1} \frac{1}{2\pi i} \int_{|w|=1} \bar{m}^{p+1}\left(\frac{1}{w}\right) f(m(w))m'(w)dw. \qquad (5.2.22)
\end{aligned}$$

We make the additional assumption on $m(w)$ that

$$m(w) = \sum_{j=1}^{q} a_j w^j, \quad a_1, a_q \neq 0, \qquad (5.2.23)$$

where $|a_2|, \ldots, |a_q|$ are all sufficiently small with respect to $|a_1|$. Then, $m(w)$ will be univalent in the unit circle $|w| \leq 1$ and hence will map it onto a simply connected *Schlicht region.* We rewrite $m^{p+1}(w)$, a polynomial of degree $s = q(p+1)$, as

$$m^{p+1}(w) = b_1 p! w^{p+1} + b_2(p+1)! w^{p+2} + \cdots + b_{s-p}(s-1)! w^s,$$

where b_j, $j = 1, \ldots, s-p$, are determined from a_j, $j = 1, \ldots, q$. By using the residue theorem, from formula (5.2.22) we obtain the following exact quadrature formula.

$$\begin{aligned}
&\iint_B \bar{z}^p f(z)dxdy \\
&= \frac{\pi}{p+1} \frac{1}{2\pi i} \int_{|w|=1} \left(\bar{b}_1 \frac{p!}{w^{p+1}} + \cdots \right. \\
&\quad \left. + \bar{b}_{s-p} \frac{(s-1)!}{w^s}\right) f(m(w))m'(w)dw \\
&= \frac{\pi}{p+1} \left[\bar{b}_1 (m'(w) f(m(w))^{(p)} |_{w=0} + \cdots \right. \\
&\quad \left. + \bar{b}_{s-p} (m'(w) f(m(w))^{(s-1)} |_{w=0}\right] \\
&= Q(D) f(0), \qquad (5.2.24)
\end{aligned}$$

where $Q(D)$ is a linear differential operator of order $s-1$ whose coefficients depend on $m(w)$ but are independent of $f(z)$.

We now consider several examples.

Example 5.2.3 If $m(w) = w + aw^2$, then $m'(w) = 1 + 2aw$, $m''(w) = 2a$, and $\bar{m}(w) = w + \bar{a}w^2$. For sufficiently small $|a|$ (for instance, $|a| < 1/2$), $m(w)$ is univalent. Denote by B the region in the z-plane that $m(w)$ maps $|w| \leq 1$ onto. From equation (5.2.24),

$$\begin{aligned}
&\iint_B f(z)dxdy \\
&= \pi \frac{1}{2\pi i}\int_{|w|=1} f(m(w))m'(w)\left(\frac{1}{w} + \frac{\bar{a}}{w^2}\right)dw \\
&= \pi\left[(1+2|a|^2)f(0) + \bar{a}f'(0)\right].
\end{aligned} \tag{5.2.25}$$

We next investigate the case when $m(w)$ is a rational function of w. Let

$$P(w) = (w-\alpha_1)(w-\alpha_2)\cdots(w-\alpha_n) \tag{5.2.26}$$

and

$$R(w) = (w-\beta_1)(w-\beta_2)\cdots(w-\beta_p), \tag{5.2.27}$$

where $0 < |\alpha_j| < 1$, $j = 1, \ldots, n$; and β_i $(i = 1, \ldots, p)$ are distinct from α_j $(j = 1, \ldots, n)$. Also let $Q(w) = w^n P(1/w)$, $T(w) = w^p R(1/w)$, and $m(w) = awT(w)/Q(w)$, $a \neq 0$. Then $m(w)$ can be written as

$$\begin{aligned}
m(w) &= \frac{awT(w)}{Q(w)} \\
&= \frac{aw(1-\beta_1 w)\cdots(1-\beta_p w)}{(1-\alpha_1 w)\cdots(1-\alpha_n w)} \\
&= aw + \cdots .
\end{aligned} \tag{5.2.28}$$

If $|\alpha_j|$ and $|\beta_k|$ $(j = 1, \ldots, n;\ k = 1, \ldots, p)$ are sufficiently small, then $m(w)$ is univalent in $|w| \leq 1$. Therefore, $m(w)$ maps $|w| \leq 1$ one-to-one conformally onto a region B in the z-plane. Noting that $m'(w) = [awT(w)/Q(w)]'$ is regular in $|w| \leq 1$ and

$$\begin{aligned}
\bar{m}\left(\frac{1}{w}\right) &= \frac{\bar{a}(1/w)(1-\bar{\beta}_1/w)\cdots(1-\bar{\beta}_p/w)}{(1-\bar{\alpha}_1/w)\cdots(1-\bar{\alpha}_n/w)} \\
&= \frac{\bar{a}w^{n-p-1}(w-\bar{\beta}_1)\cdots(w-\bar{\beta}_p)}{(w-\bar{\alpha}_1)\cdots(w-\bar{\alpha}_n)} \\
&= \frac{\bar{a}w^{n-p-1}\bar{R}(w)}{\bar{P}(w)},
\end{aligned}$$

we immediately derive that $\bar{m}(w)$ has poles at $\bar{\alpha}_1, \ldots, \bar{\alpha}_n$. It has no other singularities if $n \geq p+1$ but has an additional pole of order $p+1-n$ at $w=0$ if $n < p+1$.

From equation (5.2.22) (for $p=0$) we obtain

$$\begin{aligned}
&\iint_B f(z)dxdy \\
&= \pi \frac{1}{2\pi i} \int_{|w|=1} f(m(w))m'(w)\bar{m}\left(\frac{1}{w}\right) dw \\
&= \pi \frac{1}{2\pi i} \int_{|w|=1} f(m(w)m'(w) \\
&\quad \times \frac{\bar{a}w^{n-p-1}(w-\bar{\beta}_1)\cdots(w-\bar{\beta}_p)}{(w-\bar{\alpha}_1)\cdots(w-\bar{\alpha}_n)} dw. \qquad (5.2.29)
\end{aligned}$$

We now consider the case when $n \geq p+1$ and all α_j are distinct. Then, from the residue theorem,

$$\begin{aligned}
&\iint_B f(z)dxdy \\
&= \pi\bar{a} \sum_{j=1}^{n} f(m(\bar{\alpha}_j))m'(\bar{\alpha}_j) \frac{\bar{\alpha}_j^{n-p-1}(\bar{\alpha}_j-\bar{\beta}_1)\cdots(\bar{\alpha}_j-\bar{\beta}_p)}{\bar{P}'(\alpha_j)}. \qquad (5.2.30)
\end{aligned}$$

If α_j has multiplicity τ_j, then the corresponding exact quadrature formula contains a differential operator of the order τ_{j-1} evaluated at $\bar{\alpha}_j$.

If $n < p+1$, then the point $\alpha = 0$ is a pole, and $f(0)$ $(m(0)=0)$ exists by itself when $n = p$ or with its higher derivatives when $n < p$.

To evaluate the higher moments $\int\int_B \bar{z}^r f(z)dxdy$, we use DRE (5.2.22) and find

$$\begin{aligned}
&\iint_B \bar{z}^r f(z)dxdy \\
&= \frac{\pi}{r+1} \frac{1}{2\pi i} \int_{|w|=1} f(m(w))m'(w)(\bar{a})^{r+1} \\
&\quad \times w^{(n-p-1)(r+1)} \frac{\left[(w-\bar{\beta}_1)\cdots(w-\bar{\beta}_p)\right]^{r+1}}{\left[(w-\bar{\alpha}_1)\cdots(w-\bar{\alpha}_n)\right]^{r+1}} dw.
\end{aligned}$$

Using the residue theorem yields

$$\iint_B \bar{z}^r f(z)dxdy = \sum_{k=1}^{n}\sum_{j=0}^{r} c_{kj} f^{(j)}(m(\bar{\alpha}_k)) + \sum_{j=0}^{r} d_j f^{(j)}(0), \qquad (5.2.31)$$

where c_{kj} and d_j are independent of f and the $d_j's$ all vanish if $n \geq p+1$.

Example 5.2.4 Let $P(w) = w - \alpha$, $|\alpha| < 1$, and $R(w) = 1$. Similar to Example 5.2.3, we have $S(w) = 1$ and $Q(w) = 1 - \alpha w$. From (5.2.28) with $a = 1$, we have

$m(w) = w/(1-\alpha w)$. Thus $m'(w) = 1/(1-\alpha w)^2$ and $\bar{m}(1/w) = 1/(w-\bar{\alpha})$. It follows that

$$\begin{aligned}\iint_B f(z)dxdy &= \frac{\pi}{2\pi i}\int_{|w|=1} f\left(\frac{w}{1-\alpha w}\right)\frac{1}{(1-\alpha w)^2}\frac{1}{w-\bar{\alpha}}dw \\ &= \frac{\pi}{(1-|\alpha|^2)^2} f\left(\frac{\bar{\alpha}}{1-|\alpha|^2}\right).\end{aligned}$$

The above quadrature formula is analogous to (5.1.30) with $n = 0$.

Example 5.2.5 Let $m(w) = w(1-\beta^2 w^2)/(1-\alpha^2 w^2)$, $0 < |\alpha| < 1$, $\beta \neq \pm\alpha$, and $n = p = 2$. Take α and β sufficiently close to 0 so that $m(w)$ is univalent in the circle $|w| \leq 1$ and maps it onto a region B. Then,

$$\begin{aligned}m'(w) &= \frac{1+\alpha^2 w^2 - 3\beta^2 w^2 + \alpha^2\beta^2 w^4}{(1-\alpha^2 w^2)^2}, \\ \bar{m}\left(\frac{1}{w}\right) &= \frac{w^2-\bar{\beta}^2}{w(w^2-\bar{\alpha}^2)},\end{aligned}$$

and we have the quadrature formula

$$\iint_B f(z)dxdy = A(f(z^*) + f(-z^*)) + Cf(0), \tag{5.2.32}$$

where $z^* = \bar{\alpha}(1-\beta^2\bar{\alpha}^2)/(1-|\alpha|^4)$, $C = (\bar{\beta}/\bar{\alpha})^2$, and

$$A = \frac{1+|\alpha|^4 - 3\beta^2\bar{\alpha}^2 + \beta^2\bar{\alpha}^2|\alpha|^4}{1-|\alpha|^4}.$$

5.3 Integral regions suitable for DREs

In this section, we will give a method for finding a region B such that for a class of functions defined on B, a given DRE or a quadrature formula over B holds exactly. Since this method is based on the so-called reproducing kernel functions, in the following we will cite some basic facts of these functions from Section 12.6 of book [12].

Let S be a set in a one or higher dimensional real or complex space; z, w, ... be points in S; and X be the complete inner product space that consists of functions defined on S.

For $z, w \in S$, if a bivariate function $K(z, w)$ satisfies the following two conditions, then $K(z, w)$ is referred to as the reproducing kernel function of the complete inner product space X.

1. For any fixed $z \in S$, $K(z, w)$, considered as a function in terms of w, is in X;

2. For all functions $f(w) \in X$ and all $z \in S$,

$$f(z) = (f(w), K(z, w))_w \tag{5.3.1}$$

holds. $(f(w), K(z, w))$ is the inner product with respect to the variable w.

Theorem 5.3.1 (Aronszajn) *A complete inner product space X has a reproducing kernel if and only if for any fixed $z \in S$, the linear functional*

$$L(f) = f(z)$$

is bounded; i.e., for all functions $f \in X$, we have

$$|L(f)| \leq C_z \parallel f \parallel \tag{5.3.2}$$

for some constant C_z independent of f and for all $f \in X$.

Proof. If $K(z, w)$ is a reproducing kernel, then equation (5.3.1) holds. By using the Schwarz inequality, we have

$$\begin{aligned} |f(z)|^2 &\leq (f(w), f(w))(K(z, w), K(z, w))_w \\ &= \parallel f \parallel^2 K(z, z). \end{aligned} \tag{5.3.3}$$

The identity results from substituting $K(z, z)$ into equation (5.3.1). Hence we obtain equation (5.3.2) with $C_z = \sqrt{K(z, z)}$.

The necessity of the theorem can be proved by using the *general Fréchet–Riesz theorem* for linear functionals of complete inner product spaces. From condition (5.3.2), we can apply the Fréchet–Riesz theorem and find a function $g_z(w) \in X$ such that $f(z) = L(f) = (f(w), g_z(w))$. Hence, $g_w(z)$ is a reproducing kernel. □

It is easy to prove that if space X has a reproducing kernel, then the kernel is unique. If $K(z, w)$ and $J(z, w)$ are two reproducing kernels of X, then

$$\begin{aligned} &\parallel K(z, w) - J(z, w) \parallel^2 \\ = &(K(z, w) - J(z, w), K(z, w) - J(z, w)) \\ = &(K(z, w) - J(z, w), K(z, w)) - (K(z, w) - J(z, w), J(z, w)) \\ = &(K(z, w) - J(z, w)) - (K(z, w) - J(z, w)) = 0. \end{aligned}$$

Theorem 5.3.2 *Assume $K(z, w)$ is the reproducing kernel function of the complete inner product space X and L is a linear functional defined on X. Then the function*

$$r(z) = \overline{L_w K(z, w)}$$

is in X, and for all $f \in X$, we have

$$L(f) = (f(z), r(z)),$$

where L_w is an operator in terms of w when it is applied to $K(z, w)$.

Proof. From the Fréchet–Riesz theorem, there is a $g(z) \in X$ such that $L(f) = (f, g)$ for all $f \in X$. It follows that $\overline{L_w K(z,w)} = \overline{(K(z,w), g(z))_z} = (g(z), K(z,w))_z = g(w)$. $\square$

Because $L^2(B)$ (B is a region in the complex plan) is a complete inner product space (see [12], p. 212), Theorem 5.3.1 shows that $L^2(B)$ possesses a reproducing kernel $K(z,w)$. In addition, if region B is bounded or can be mapped one-to-one conformally onto a bounded region, then $L^2(B)$ is also a *Hilbert space.* Suppose $\{\zeta_n(z)\}_{n=1}^{\infty}$ is a complete orthonormal sequence in $L^2(B)$. From the properties of orthonormal sequences, the reproducing kernel $K(z,w)$ has the representation

$$K(z,w) = \sum_{n=1}^{\infty} \overline{\zeta_n(z)} \zeta_n(w).$$

In general, this type of kernel is called the Bergman kernel for the region B.

Theorem 5.3.3 *Suppose that $t = m(z)$ maps region B one-to-one conformally onto the unit circle C: $|t| \le 1$. Then $L^2(B)$ possesses the reproducing kernel*

$$K(z,w) = \frac{\overline{m'(z)} m'(w)}{\pi(1 - \overline{m(z)} m(w))^2}, \qquad z, w \in B. \tag{5.3.4}$$

Proof. Denote the inverse of m by $z = M(t)$. Then $m(M(t)) = t$ and $m'(M(t))M'(t) \equiv 1$. For any $f, g \in L^2(B)$, applying suitable transforms, we obtain

$$\begin{aligned} (f,g) &= \iint_B f(z)\overline{g(z)} dxdy \\ &= \iint_C f(M(t))\overline{g(M(t))} \left|M'(t)\right|^2 dudv, \end{aligned}$$

where $t = u + iv$.

We now prove that $\{u_n(z) = \sqrt{\frac{n+1}{\pi}}(m(z))^n m'(z), n = 0, 1, \ldots\}$, is a complete and orthonormal sequence in $L^2(B)$. We have

$$\begin{aligned} &\sqrt{\frac{\pi}{n+1}}(f, u_n) \\ = &\iint_B f(z)\overline{(m(z))^n m'(z)} dxdy \\ = &\iint_C f(M(t))\overline{t^n m'(M(t))} \left|M'(t)\right|^2 dudv \\ = &\iint_C f(M(t))M'(t)\overline{t^n} dudv. \end{aligned}$$

From the completeness of $1, t, t^2, \ldots$ in $L^2(C)$, we immediately know that $(f, u_n) = 0$, $n = 0, 1, \ldots$, implies $f(M(t))M'(t) \equiv 0$. Since $M' \neq 0$, we obtain

$f \equiv 0$. It follows that $\{u_n\}$ is complete for $L^2(B)$. The proof of orthonormality is similar.

As we mentioned before, the Bergman kernel of $L^2(B)$ can be presented in terms of the complete orthonormal sequence $\{u_n\}$ as

$$K(z, w) = \frac{1}{\pi}\sum_{n=0}^{\infty}(n+1)\overline{(m(z))^n m'(z)}(m(w))^n m'(w). \tag{5.3.5}$$

An application of the identity

$$(1-x)^{-2} = \sum_{n=0}^{\infty}(n+1)x^n$$

to equation (5.3.5) yields expression (5.3.4). □

We will use the above results on the reproducing kernel functions to solve the following problem. For a given linear functional L, possibly an integro-differential functional, applicable to a class X of analytic functions, we will try to find a region B in the z-plane such that L has a representation as a double integral over B. That is, find a region B such that

$$\iint_B f(z)dxdy = L(f) \tag{5.3.6}$$

holds exactly for all $f \in X$. It simplifies our thinking if we take X and L to be the complete inner product space $L^2(B)$ and the bounded linear functional, respectively. However, even with these assumptions, the problem cannot always be solved. For example, if $L(f) = f'(0) = \int\int_B f(z)dxdy$, we obtain a contradiction by setting $f \equiv 1$. Thus, the method we give below is only for problems that have solutions. In the course of the discussion, we will also attempt to answer some interesting and significant questions raised in [15]: how to characterize those functionals for which there is a solution to (5.3.6)? If there is a solution, is the solution unique in some sense? Is the solution unique if the region B is restricted to being simply connected?

Suppose $L^2(B)$ possesses the Bergman kernel $K(z, w)$. From Theorem 5.3.2, we can rewrite $L(f)$ as

$$L(f) = \iint_B f(z)L_w K(z, w)dxdy, \quad f \in L^2(B). \tag{5.3.7}$$

To equate (5.3.6) and (5.3.7), we set

$$\overline{L_w K(z, w)} \equiv 1. \tag{5.3.8}$$

This is a necessary and sufficient condition for (5.3.6). Our problem is thus equivalent to the following: Find, if possible, a region B for which (5.3.8) holds. Assume that B is a simply connected region. Combining (5.3.4) and (5.3.8) gives

$$\frac{1}{\pi}\overline{L_w\left(\frac{\overline{m'(z)}m'(w)}{(1-\overline{m(z)}m(w))^2}\right)} = 1, \tag{5.3.9}$$

where $t = m(z)$ maps the region B one-to-one conformally onto the unit circle $|t| \le 1$. Equation (5.3.9) is a functional-differential equation of $m(z)$, which, as we pointed out, is not always solvable. However, it is useful in applications after further reduction. An application of expression (5.3.5) yields

$$\frac{1}{\pi}\sum_{n=0}^{\infty}(n+1)(m(z))^n m'(z)\overline{L\left((m(w))^n m'(w)\right)} = 1. \tag{5.3.10}$$

If $|z|, |w| \le 1-\delta, 0 < \delta \le 1$, then the series in (5.3.10) converges absolutely and uniformly.

$$\frac{1}{\pi}\frac{d}{dz}\sum_{n=0}^{\infty}(m(z))^{n+1}\,\overline{L\left((m(w))^n m'(w)\right)} = 1. \tag{5.3.11}$$

Taking an integral on each side of the above equation from $z = 0$ to z in $|z| < 1$, we obtain

$$m(z)\sum_{n=0}^{\infty}(m(z))^n\,\overline{L\left((m(w))^n m'(w)\right)} = \pi z. \tag{5.3.12}$$

It follows that

$$\begin{aligned} \pi\bar{z} &= \overline{m(z)}L_w\sum_{n=0}^{\infty}\overline{(m(z))^n}\,(m(w))^n\,m'(w) \\ &= \overline{m(z)}L_w\left(\frac{m'(w)}{1-m(w)\overline{m(z)}}\right), \end{aligned} \tag{5.3.13}$$

or equivalently,

$$\pi\bar{z} = -\overline{m(z)}L_w\frac{d}{dw}\frac{1}{\overline{m(z)}}\log\left(1-m(w)\overline{m(z)}\right).$$

Finally, we obtain

$$-\pi\bar{z} = L_w\frac{d}{dw}\log\left(1-m(w)\overline{m(z)}\right). \tag{5.3.14}$$

We now give several examples of applications.

Example 5.3.1 Let $L(f) = \pi a^2 f(0)$. Then, from (5.3.12),

$$m(z)\pi a^2 \overline{m'(0)} = \pi z.$$

Differentiating the above equation at the point $z = 0$ gives

$$\pi = \pi a^2 |m'(0)|^2.$$

It follows that $m'(0) = e^{i\psi}/a$. Hence, $z = am(z)e^{-i\psi}$ and $m(z) = ze^{i\psi}/a$. Therefore, B is the circle $|z| \le a$, and (5.1.30) with $n = 0$ is produced.

Example 5.3.2 With the functional $L(f) = \int_{-1}^{1} f(x)dx$, DRE (5.3.6) can be written as

$$\iint_B f(z)dxdy = \int_{-1}^{1} f(x)dx \tag{5.3.15}$$

for a suitable region B, if it exists, and for all $f \in L^2(B)$, while equation (5.3.14) becomes

$$\begin{aligned} -\pi\bar{z} &= \int_{-1}^{1} \frac{d}{dw} \log(1 - m(w)\overline{m(z)})dw \\ &= \log(1 - m(1)\overline{m(z)}) - \log(1 - m(-1)\overline{m(z)}) \\ &= \log\left(\frac{1 - m(1)\overline{m(z)}}{1 - m(-1)m(z)}\right). \end{aligned} \tag{5.3.16}$$

To obtain symmetry we assume

$$m(0) = 0, \quad m(-1) = -m(1) = \alpha. \tag{5.3.17}$$

Substituting $z = 1$ and $z = -1$ into (5.3.16) produces the two identical equations

$$e^{-\pi} = (1 - |\alpha|^2)/(1 + |\alpha|^2); \quad e^{\pi} = (1 + |\alpha|^2)/(1 - |\alpha|^2). \tag{5.3.18}$$

The solution is thus

$$|\alpha|^2 = (e^{\pi} - 1)/(e^{\pi} + 1) = (1 - e^{-\pi})/(1 + e^{-\pi}), \tag{5.3.19}$$

or

$$\alpha = e^{i\theta}\sqrt{\frac{1 - e^{-\pi}}{1 + e^{-\pi}}}, \quad 0 \le \theta \le 2\pi. \tag{5.3.20}$$

We select $\theta = 0$, then $\alpha \approx 0.958 < 1$. With this value for α, from (5.3.16)–(5.3.18), we obtain a mapping function

$$t = m(z) = \frac{1}{\alpha}\left(\frac{1 - e^{\pi z}}{1 + e^{\pi z}}\right) = -\frac{1}{\alpha}\tanh\frac{\pi}{2}z, \tag{5.3.21}$$

while the region B that is suitable for DRE (5.3.15) can be given by the inverse mapping of expression (5.3.21):

$$z = M(t) = \frac{1}{\pi}\log\left(\frac{1-\alpha t}{1+\alpha t}\right) = -\frac{2}{\pi}arctan\,h(\alpha t), \tag{5.3.22}$$

where $\alpha \approx 0.958$. As t traces the unit circle, $\zeta = (1-\alpha t)/(1+\alpha t)$ traces a circle that lies in $Re\,\zeta > 0$ and has the center $((1+\alpha^2)/(1-\alpha^2), 0)$ and the radius $2\alpha/(1-\alpha^2)$. From equation (5.3.22), the image in the z plane of $|t| = 1$ is therefore Schlicht. In addition, the segment $[-1, 1]$ is interior to the image because $|m(\pm 1)| = \alpha < 1$, and $m(z)$ is real for real z. Therefore, the function in (5.3.22) maps the unit circle onto a simply connected region B that satisfies (5.3.15). In addition, B is an ellipse-like figure having a semimajor axis $a = \log((1+\alpha)/(1-\alpha))/\pi \approx 1.22$ and a semiminor axis $b = arctan(2\alpha/(1-\alpha^2))/\pi \approx 0.486$.

We now find the Schwarz function. From (5.3.21), (5.3.22), and the expression of the Schwarz function (5.1.25), we obtain

$$S(z) = \bar{m}\left(\frac{1}{M(z)}\right) = \frac{1}{\pi}\log\left(\frac{1-e^{\pi(z+1)}}{e^{\pi} - e^{\pi z}}\right). \tag{5.3.23}$$

$S(z)$ possesses logarithmic singularities at $z = \pm 1, \pm 2ki$, $k = 0, 1, \ldots$, $z = \pm 1$ being the only singularities in B. By making a cut along the real axis from $z = 1$ to $z = -1$, we can define the single-valued branch, $S_u(z)$, of $S(z)$ along the upper edge of the cut and the branch $S_l(z)$ along the lower edge of the cut as follows.

$$\begin{aligned} S_u(z) &= \frac{1}{\pi}\log\left(\frac{e^{\pi(x+1)}-1}{e^{\pi} - e^{\pi x}}\right) - i \\ S_l(z) &= \frac{1}{\pi}\log\left(\frac{e^{\pi(x+1)}-1}{e^{\pi} - e^{\pi x}}\right) + i, \end{aligned} \tag{5.3.24}$$

where $-1 \le x \le 1$. From (5.1.27), we have

$$\iint_B f(z)dxdy = \frac{1}{2i}\int_{\partial B} S(z)f(z)dz.$$

Furthermore, replacing ∂B by a circuit consisting of $-(1-\epsilon) \le x \le 1-\epsilon$ augmented by two circles of radius ϵ at $x = 1$ and $x = -1$, we obtain

$$\iint_B f(z)dxdy = \frac{1}{2i}\int_{-1}^{1}(S_l(x)f(x) - S_u(x)f(x))dx. \tag{5.3.25}$$

Here, the limiting process is valid because $\lim_{\epsilon\to 0}\epsilon\log\epsilon = 0$. Thus, by substituting $S_u(z)$ and $S_l(z)$ given in (5.3.24) into (5.3.25), we establish the equation (5.3.15).

Similarly, we have the following DREs for the higher moments.

$$\begin{aligned}\iint_B \bar{z}^p f(z)dxdy &= \frac{1}{2i(p+1)}\int_{\partial B} S^{p+1}(z)f(z)dz \\ &= \frac{1}{2i(p+1)}\int_{-1}^{1}\left(S_l^{p+1}(x) - S_u^{p+1}(x)\right)f(x)dx \\ &= \frac{1}{2i(p+1)}\int_{-1}^{1}\zeta_p(x)f(x)dx, \qquad (5.3.26)\end{aligned}$$

where

$$\begin{aligned}&\zeta_p(x) \\ =\ &\left(\frac{1}{\pi}\log\left(\frac{e^{\pi(x+1)}-1}{e^{\pi}-e^{\pi x}}\right)+i\right)^{p+1} - \left(\frac{1}{\pi}\log\left(\frac{e^{\pi(x+1)}-1}{e^{\pi}-e^{\pi x}}\right)-i\right)^{p+1} \\ =\ &2iIm\left(\frac{1}{\pi}\log\left(\frac{e^{\pi(x+1)}-1}{e^{\pi}-e^{\pi x}}\right)+i\right)^{p+1}. \qquad (5.3.27)\end{aligned}$$

In particular, if $p = 1$ and $p = 2$, we have

$$\iint_B \bar{z}f(z)dxdy = \frac{1}{\pi}\int_{-1}^{1}\log\left(\frac{e^{\pi(x+1)}-1}{e^{\pi}-e^{\pi x}}\right)f(x)dx \qquad (5.3.28)$$

and

$$\begin{aligned}&\iint_B \bar{z}^2 f(z)dxdy \\ =\ &\frac{1}{\pi^2}\int_{-1}^{1}\log^2\left(\frac{e^{\pi(x+1)}-1}{e^{\pi}-e^{\pi x}}\right)f(x)dx - \frac{1}{3}\int_{-1}^{1}f(x)dx, \qquad (5.3.29)\end{aligned}$$

respectively.

5.4 Additional topics

Examples 5.1.2 and 5.3.2 (see equations (5.1.37) and (5.3.23)) show how, if the Schwarz function of region B is regular in a slit B, then the integral $\int\int_B f dxdy$ can be reduced to one of the form $\int_{-1}^{1}\mu(x)f(x)dx$. In this section, we will use the method of "opening up the slit" to construct general regions B with the property. Then we will give applications of DREs in Fourier expansion.

Let w_s be the w plane with two slits $[1/\alpha, \infty)$ and $(-\infty, -1/\alpha]$ along the real axis. Here, $0 < \alpha < 1$. The mapping

$$u = \Gamma(w) = \frac{1-\sqrt{1-\alpha^2 w^2}}{\alpha w} \qquad (5.4.1)$$

maps w_s one-to-one conformally onto the circle $|u| \leq 1$ in the u plane. Hence,

$$w = \gamma(u) = \frac{2u}{\alpha(u^2+1)}. \tag{5.4.2}$$

For $u = e^{i\theta}$, the above expression can be written as

$$w = \frac{2}{\alpha(u+u^{-1})} = \frac{1}{\alpha \cos\theta}. \tag{5.4.3}$$

w goes from ∞ to $1/\alpha$ and back to ∞ along the right slit when $-\pi/2 \leq \theta \leq \pi/2$; it goes from $-\infty$ to $-1/\alpha$ then back to $-\infty$ along the left slit when $\pi/2 \leq \theta \leq 3\pi/2$.

Denote the image of $|w| = |e^{i\xi}| = 1$ in the u plane by C. Obviously, C is contained in $|u| \leq 1$ and has the equation

$$C: \ u = \frac{1-\sqrt{1-\alpha^2 e^{2i\xi}}}{\alpha e^{i\xi}}. \tag{5.4.4}$$

In addition, from (5.4.1) the image in the u plane of $w = \pm\alpha$ is $\pm\beta$, $\beta = (1-\sqrt{1-\alpha^4})/\alpha^2$.

We now introduce an arbitrary analytic function $H(u)$ subject to the following restrictions. We must point out that some of the conditions are not absolutely necessary and are included just to simplify the situation.

(1) The function $z = H(u) = a_1 u + a_2 u^2 + \cdots$ is regular in $|u| < 1$ and has real coefficients.

(2) $H(u)$ is continuous in $|u| \leq 1$ or mildly singular on $|u| = 1$.

(3) $H(u)$ is univalent in and on C with the property $H(-\beta) = a < b = H(\beta)$.

Denote by B the image in the z plane of C under the function H. Then B contains the segment $[a, b]$ of the real axis. In addition, the inverse function of H, denoted by h, performs a one-to-one conformal map of B onto C.

Therefore, the function

$$z = m(w) \equiv H(\Gamma(w)) \tag{5.4.5}$$

is regular in w and is univalent in $|w| \leq 1$ (because the image of $|w| = 1$ under Γ is C.) It follows that $z = m(w)$ maps $|w| \leq 1$ one-to-one conformally onto B and the boundary of B, ∂B, is an analytic curve.

Denote by B_s the region B with a slit $[a, b]$ along the real axis and by C_s the region that is bounded by C and has the slit $[-\beta, \beta]$ removed. We can prove that $S(z)$, the Schwarz function for ∂B, is regular in B_s. Since the inverse of $z = m(w)$ is $w = M(z) = \gamma(h(z))$ and $\bar{m} = m$, we have

$$S(z) = H(\Gamma(1/\gamma(h(z))). \tag{5.4.6}$$

Therefore, $h(z)$ is regular in B_s and takes values in C_s. $\gamma(h(z))$ is therefore regular in B_s and takes values in $|w| \leq 1$ except for the interval $[-\alpha, \alpha]$ along the real axis. Hence, $1/\gamma(h(z))$ is regular in B_s and takes values in $w_s - (|w| \leq 1)$. It follows that $\Gamma(1/\gamma(h(z)))$ is regular in B_s and takes values in $(|u| \leq 1) - C$. Therefore, finally, $S(z) = H(\Gamma(1/\gamma(h(z)))$ is regular in B_s.

Since $H(u)$ is continuous (or has mild singularity) on $|u| = 1$, $S(z)$ is continuous from the interior of B_s on the upper and lower edges of the slit. We now find $S_u(x)$ and $S_l(x)$, the Schwarz functions for the upper edge and the lower edge respectively.

We have $z = x + iy$. As x goes from a to b and back to a, $h(x)$ goes from $-\beta$ to β to $-\beta$ and $\gamma(h(x))$ from $-\alpha$ to α to $-\alpha$. Therefore, $1/(\gamma(h(x))$ goes from $-1/\alpha$ to $-\infty$, ∞ to $1/\alpha$, $1/\alpha$ to ∞, $-\infty$ to $-1/\alpha$. For $u = e^{i\theta}$, $w = 1/\alpha \cos\theta$ and $\Gamma(1/\gamma(h(x))) = e^{i\theta}$. Here, θ goes from $-\pi$ to 0 as x runs from a to b and from 0 to π as x runs from b to a. Thus we can write

$$\begin{aligned} S_u(x) &= H(e^{i\theta}), \quad 0 \leq \theta \leq \pi, \\ S_l(x) &= H(e^{i\theta}), \quad -\pi \leq \theta \leq 0. \end{aligned} \tag{5.4.7}$$

Since $S(z)$ is regular in B and continuous or mildly singular on $[a, b]$, we may apply DRE (5.1.27) for $n = 0$ and collapse the contour ∂B to $[a, b]$ to obtain

$$\begin{aligned} \iint_B f(z)dxdy &= \frac{1}{2i}\int_{\partial B} S(z)f(z)dz \\ &= \frac{1}{2i}\int_a^b (S_l(x) - S_u(x))\, f(x)dx. \end{aligned} \tag{5.4.8}$$

If $H(u)$ is an odd function, then $S_l(x) = H(e^{i\theta})$ and $S_u(x) = H(e^{-i\theta})$. Therefore, $S_l(x) - S_u(x) = 2i Im\, H(e^{i\theta})$ and

$$\iint_B f(z)dxdy = \int_{-b}^b Im\, H(e^{i\theta(x)})f(x)dx, \tag{5.4.9}$$

where $\theta = \theta(x)$ is determined from

$$\cos\theta = \alpha^2\frac{h^2(x)+1}{2h(x)}. \tag{5.4.10}$$

In addition, if H is odd and is real on the real axis, then we have the DRE formula

$$\begin{aligned} &\iint_B \bar{z}^p f(z)dxdy \\ &= \frac{1}{2i(p+1)}\int_{-b}^b \left(H^{p+1}(e^{i\theta}) - H^{p+1}(e^{-i\theta})\right) f(x)dx. \end{aligned} \tag{5.4.11}$$

Since $H^{p+1}(e^{-i\theta}) = \overline{H^{p+1}(e^{i\theta})}$,

$$\iint_B \bar{z}^p f(z)dxdy = \frac{1}{p+1}\int_{-b}^{b} Im\, H^{p+1}(e^{i\theta})f(x)dx. \tag{5.4.12}$$

Example 5.4.6 Assume that $z = H(u) = u$. Then this function obviously has the required properties and has the inverse function $u = h(z) = z$. Moreover it is odd. Hence, we have the corresponding

$$\beta = b = \frac{1-\sqrt{1-a^4}}{a^2}.$$

The corresponding region B is identical to the region bounded by C. Since $1/\gamma(h(x)) = \alpha(x^2+1)/2x$, we have $\cos\theta = \alpha^2(x^2+1)/2x$ and

$$\sin\theta = \frac{\sqrt{(\alpha^2(x^2+1)+2x)(\alpha^2(x^2+1)-2x)}}{\alpha^2(x^2+1)} \equiv \zeta(x).$$

Moreover, $Im\, H(e^{i\theta}) = \sin\theta$ so that the DRE formula

$$\iint_B f(z)dxdy = \int_{-\beta}^{\beta} \zeta(x)f(x)dx$$

holds.

At the end of this section, we discuss Fourier expansion using DRE formulas. Suppose that B is a bounded, simply connected region of the complex plane, whose boundary ∂B is an analytic curve. There exists a complete orthonormal set of polynomials $\{p_n^*(z)\}$ for the Hilbert space $L^2(B)$, where the degree of $p_n^*(z)$ is n. Hence, for any $f(z) \in L^2(z)$, we have the *Fourier expansion*

$$\begin{aligned} f(z) &= \sum_{n=0}^{\infty}\langle f, p_n^*\rangle p_n^*(z) \\ &= \sum_{n=0}^{\infty}\left(\iint_B f(z)\overline{p_n^*(z)}dxdy\right) p_n^*(z). \end{aligned} \tag{5.4.13}$$

The above series converges uniformly and absolutely in compact subregions of B. Write

$$p_n^*(z) = \alpha_{nn}z^n + \cdots + \alpha_{n1}z + \alpha_{n0}, \quad \alpha_{nn} \neq 0; \tag{5.4.14}$$

then, for certain regions B, the Fourier coefficients $\int\int_B f\overline{p_n^*}dxdy$ have alternate DRE as linear differential operators or real integrals operating on f. For example, if B is the bicircular quartic Q, then from (5.2.7) we have

$$\iint_Q \bar{z}^n f(z)dxdy = R_n(D)f(\epsilon) + S_n(D)f(-\epsilon), \tag{5.4.15}$$

where $R_n(D)$ and $S_n(D)$ are certain nth order operators independent of f. From (5.4.14), we obtain

$$\begin{aligned}\langle f, p_n^*\rangle &= \iint_Q f(z)\overline{p_n^*(z)}dxdy \\ &= R_n^*(D)f(\epsilon) + S_n^*(D)f(-\epsilon), \end{aligned} \tag{5.4.16}$$

where $R_n^*(D)$ and $S_n^*(D)$ are operators obtained from (5.4.15) by substituting (5.4.14) into the integral of (5.4.16). The Fourier expansion (5.4.13) can be written as

$$f(z) = \sum_{n=0}^{\infty} \left(R_n^*(D)f(\epsilon) + S_n^*(D)f(-\epsilon)\right) p_n^*(z). \tag{5.4.17}$$

Substituting $f(z) = p_m^*(z)$ into the above expression, from the orthonormality of $\{p_n^*(z)\}$, we have the biorthonormality

$$R_n^*(D)p_m^*(\epsilon) + S_n^*(D)p_m^*(-\epsilon) = \delta_{nm}. \tag{5.4.18}$$

Define an operator T_k by $T_k g(z) = R_k^*(D)g(\epsilon) + S_k^*(D)g(-\epsilon)$ and denote

$$f_N(z) = \sum_{n=0}^{N} \langle f, p_n^*\rangle p_n^*(z).$$

Then, by applying the biorthonormality property (5.4.18), we obtain the interpolation to f,

$$T_k f_N(z) = T_k f(z), \quad k = 0, 1, \dots, N. \tag{5.4.19}$$

In addition, (5.4.18) implies *Parseval's equality* $\|f\|^2 = \sum_{n=0}^{\infty} |T_n(f)|^2$. Furthermore, the biorthonormality property (5.4.18) and the interpolation (5.4.19) can also be used to solve the minimum problem

$$I = \min_{p\in P_N} \iint_Q |f(z) - p(z)|^2 dxdy, \tag{5.4.20}$$

where P_N designates the class of all polynomials of degree $\le N$. In fact, the minimizing p is

$$p(z) = \sum_{n=0}^{N} \langle f, p_n^*\rangle p_n^*(z) = \sum_{n=0}^{N} T_n(f) p_n^*(z),$$

which satisfies

$$R_n(D)p(\epsilon) + S_n(D)p(-\epsilon) = R_n(D)f(\epsilon) + S_n(D)f(-\epsilon)$$

for $n = 0, 1, \ldots, N$. The minimum value is given by

$$I = \iint_Q |f|^2 dxdy - \sum_{n=0}^{N} |T_n(f)|^2 = \sum_{n=N+1}^{\infty} |T_n(f)|^2.$$

It should be pointed out that a formal series of the type in (5.4.17) can be defined for classes of functions much wider than $L^2(Q)$; it is sufficient for f to have all derivatives at $z = \pm\epsilon$. However, the convergence, summability, etc. of such generalized Fourier series need to be investigated. In addition, similar remarks apply to other special regions.

Secondly, a region may have associated with it several different sets of orthonormal polynomials, which may or may not be of the form (5.4.14).

Finally, the techniques of integration with wavelet functions shown in Section 2.3 may be applied to some special regions associated with wavelet functions as the orthonormal functions over the region.

Chapter 6

Exact DREs Associated With Differential Equations

With some restrictions on the integrands, we can construct a type of exact DRE, or DREs with zero remainders.

Differential equations and DREs have a natural connection. Inspired by Ghizzetti and Ossicini's work [22] on the quadrature formulas related to the solutions of ordinary differential equations, Kratz constructed a type of DRE by using Green's theorem and the solutions of partial differential equations. The exact DREs associated with the solutions of ordinary differential equations and partial differential equations will be discussed in Sections 1 and 2 respectively. Sections 3 and 4 will give applications of Kratz's exact DREs in constructing optimal BTQFs and in the boundary element method.

6.1 DREs and ordinary differential equations

In this chapter, we will generalize Kratz's result to a higher dimensional setting. We begin by taking a look at results from Ghizzetti and Ossicini's book [22].

Let $AC^k[a,b]$ be the class of all functions $f(x)$ whose kth derivative $f^{(k)}(x)$ is *absolutely continuous* in $[a,b]$, $k = 0, 1, \ldots$. It is known that $f(x) \in AC^k[a,b]$ implies that $f^{(k+1)}(x)$ exists almost everywhere in $[a,b]$ and that $f^{(k+1)}(x) \in L[a,b]$. Here, $L[a,b]$ is the class of all Lebesgue integrable functions. Following [22], we consider the differential operator

$$E = \sum_{k=0}^{n} a_k(x) D^{(n-k)}.$$

Here $D^{(m)} = d^m/dx^m$, $m = 1, 2, \ldots$, $D^{(1)} \equiv D$, and the coefficients $a_k(x)$ satisfy $a_0(x) = 1$, $a_k(x) \in AC^{n-k-1}$ for $k = 1, 2, \ldots$, and $a_n(x) \in L[a,b]$.

Hence, for any $f(x) \in AC^{n-1}[a, b]$, $(Eu)(x)$ exists almost everywhere in $[a, b]$ and is in $L[a, b]$. Define the reduced operators associated with the operator E as

$$E_r = \sum_{k=0}^{r} a_k(x) D^{r-k}, \quad r = 0, 1, \dots, n-1.$$

Obviously $E_n = E$. With the same coefficients $a_k(x)$ of operator E, we define its *adjoint operator* E^* by $(E^*v)(x) = \sum_{k=0}^{n}(-1)^{n-k} D^{n-k}[a_k(x)v(x)]$. Hence, the operator E^* can be written as

$$E^* = \sum_{k=0}^{n} a_k^*(x) D^{n-k},$$

where

$$a_k^*(x) = \sum_{j=0}^{k} (-1)^{n-j} \binom{n-j}{k-j} a_j^{(k-j)}(x)$$

for $k = 0, 1, \dots, n$. Noting the conditions that $a_k(x)$ satisfy, we immediately know that $a_0^*(x) = (-1)^n$, $a_k^*(x) \in AC^{n-k-1}[a, b]$ for $k = 1, 2, \cdot, n-1$, and $a_n^*(x) \in L[a, b]$.

Hence, if $g(x) \in AC^{n-1}[a, b]$, then $E^*(g)(x)$ exists almost everywhere in $[a, b]$ and is in $L[a, b]$. Similarly, we can define the reduced operators associated with E^* as

$$(E_r^* v)(x) = \sum_{k=0}^{r} (-1)^{r-k} D^{r-k}[a_k(x)v(x)], \quad r = 1, 2, \dots, n.$$

From the above definition, we have the recurrence formula

$$(E_r^* v)(x) = -D(E_{r-1}^* v)(x) + a_r(x)v(x), \quad r = 1, 2, \dots, n.$$

Obviously, $E^{**} = E$.

For any $u(x), v(x) \in AC^{k-1}[a, b]$, $k = 1, 2, \dots$, using mathematical induction, one may prove the relation

$$\begin{aligned} & v(x)u^{(k)}(x) - (-1)^k u(x) v^{(k)}(x) \\ = \; & D \sum_{j=0}^{k-1} (-1)^{k-j-1} u^{(j)}(x) v^{(k-j-1)}(x). \end{aligned}$$

Furthermore, from the definition of the operators E, E^*, E_r^* and the properties of $a_k(x)$, we obtain the *Green–Lagrange identity* that holds almost everywhere in $[a, b]$ for every pair of functions $u(x), v(x) \in AC^{n-1}[a, b]$,

$$\begin{aligned} & v(x)(Eu)(x) - u(x)(E^*v)(x) \\ = \; & D \sum_{k=0}^{n-1} u^{(k)}(x)(E_{n-k-1}^* v)(x). \end{aligned} \tag{6.1.1}$$

Consider the *nonhomogeneous linear differential equation*

$$(Eu)(x) = f(x), \quad x \in [a, b], \tag{6.1.2}$$

where $f(x) \in L[a, b]$. Denote by $\{u_1(x), \dots, u_n(x)\}$, $u_i(x) \in AC^{n-1}[a, b]$, a *fundamental system of solutions* of the *homogeneous equation* $(Eu)(x) = 0$, and let $U(x) \neq 0$ be the corresponding *Wronskian*. Then, the general solution of equation (6.1.2) is

$$u(x) = \sum_{j=1}^{n} \alpha_j u_j(x) + \int_a^x K(x, \xi) f(\xi) d\xi, \tag{6.1.3}$$

where $\alpha_1, \dots, \alpha_n$ are arbitrary constants and $K(x, \xi)$ is the *Cauchy resolvent kernel*; i.e., $K(x, \xi)$ is the particular solution of the homogeneous equation $(Eu)(x) = 0$ that satisfies the initial conditions

$$\left[\frac{\partial^k}{\partial x^k} K(x, \xi)\right]_{x=\xi} = \delta_{k,n-1}$$

for $k = 0, 1, \dots, n-1$ at the point ξ. Here, $\delta_{r,s}$ is the Kronecker function. We can express $K(x, \xi)$ as

$$is K(x, \xi) = \sum_{i=1}^{n} u_i(x) v_i(\xi), \tag{6.1.4}$$

where $v_i(x)$, $i = 1, \dots, n$, are determined by the system of linear equations

$$\sum_{i=1}^{n} u_i^{(k)}(x) v_i(x) = \delta_{k,n-1}, \quad k = 0, 1, \dots, n-1. \tag{6.1.5}$$

The determinant of the above system is $U(x)$. Hence, the solution (6.1.3) of the equation $(Eu)(x) = f(x)$ can be rewritten as

$$u(x) = \sum_{i=1}^{n} u_i(x) \left[\alpha_i + \int_a^x v_i(\xi) f(\xi) d\xi\right].$$

It can also be proved that the solution of system (6.1.5), $\{v_1(x), \dots, v_n(x)\}$, is a fundamental system of solutions of the homogeneous adjoint equation $(E^* v)(x) = 0$ and of the Wronskian $V(x) = 1/U(x)$. Similarly, the general solution of the nonhomogeneous $(E^* v)(x) = g(x)$, $g(x) \in L[a, b]$, is

$$\begin{aligned} v(x) &= \sum_{i=1}^{n} \beta_i v_i(x) + \int_a^x K^*(x, \xi) g(\xi) d\xi \\ &= \sum_{i=1}^{n} v_i(x) \left[\beta_i - \int_a^x u_i(\xi) g(\xi) d\xi\right]. \end{aligned} \tag{6.1.6}$$

Here β_i, $i = 1, \ldots, n$, are arbitrary constants, and $K^*(x, \xi)$ is the corresponding Cauchy resolvent kernel with respect to E^*, which can be expressed as (see (6.1.4))

$$K^*(x, \xi) = -K(\xi, x) = -\sum_{i=1}^{n} v_i(x) u_i(\xi). \tag{6.1.7}$$

We now use the solutions of a differential equation to construct the quadrature formula of the integral

$$\int_a^b g(x)u(x)dx, \tag{6.1.8}$$

where $g(x) \in L[a, b]$ is a weight function assumed to be nonzero on a set of positive measure and $u(x) \in AC^{n-1}[a, b]$. Let

$$a = x_0 \le x_1 \le \cdots \le x_m \le x_{m+1} = b$$

be a partition of the interval $[a, b]$. Here, the x_i $(i = 0, \ldots, m + 1)$ are called the nodes fixed in $[a, b]$. For a differential operator E, we call the quadrature formula

$$\int_a^b g(x)u(x)dx = \sum_{j=0}^{n-1} \sum_{i=1}^{m} A_{i,j} u^{(j)}(x_i) + R[u(x)] \tag{6.1.9}$$

the elementary quadrature formula of (6.1.8) relative to the nodes $x_1, \ldots, x_m$ and to the linear operator E. The coefficients $A_{i,j}$ are independent of $u(x)$ and the remainder, a linear functional of $u(x)$, vanishes if $u(x)$ is a solution of the homogeneous linear differential equation $(Eu)(x) = 0$.

We now discuss the construction of the elementary quadrature formula of integral (6.1.8) relative to the nodes $x_1, \ldots, x_m$ and to the linear operator E. Assume that $\phi_1(x), \ldots, \phi_{m-1}(x) \in AC^{n-1}[a, b]$ are arbitrary $m - 1$ solutions of the nonhomogeneous linear differential equation $(E^*\phi)(x) = g(x)$, and that the equation also has two other solutions $\phi_0(x)$ and $\phi_m(x)$ that satisfy the initial conditions $\phi_0^{(k)}(a) = 0$ and $\phi_m^{(k)}(b) = 0$ $(k = 0, 1, \ldots, n - 1)$, respectively. Using equations (6.1.6) and (6.1.7) yields

$$\phi_0(x) = -\int_a^x K(\xi, x)g(\xi)d\xi, \quad \phi_m(x) = \int_x^b K(\xi, x)g(\xi)d\xi. \tag{6.1.10}$$

Substituting $v(x) = \phi_i(x)$, $i = 0, 1, \ldots, m$, and noting $(E^*\phi_i)(x) = g(x)$ we obtain

$$\phi_i(x)(Eu)(x) - g(x)u(x) = D \sum_{k=0}^{n-1} u^{(k)}(x)(E^*_{n-k-1}\phi_i)(x)$$

for $i = 0, 1, \ldots, m$. Integrating on both sides of the above equation,

$$\begin{aligned}\int_a^b g(x)u(x)dx &= -\sum_{i=0}^{m}\left[\sum_{k=0}^{n-1} u^{(k)}(x)(E_{n-k-1}^{*}\phi_i)(x)\right]_{x_i}^{x_{i+1}} \\ &\quad +\sum_{i=0}^{m}\int_{x_i}^{x_{i+1}} \phi_i(x)(Eu)(x)dx. \qquad (6.1.11)\end{aligned}$$

Because $\phi_0^{(k)}(a) = 0$ and $\phi_m^{(k)}(b) = 0$,

$$\left[(E_{n-k-1}^{*}\phi_0)(x)\right]_{x=x_0} = 0, \quad \left[(E_{n-k-1}^{*}\phi_m)(x)\right]_{x=x_{m+1}} = 0$$

for $k = 0, 1, \ldots, n-1$.

Hence,

$$\begin{aligned}&-\sum_{i=0}^{m}\left[\sum_{k=0}^{n-1} u^{(k)}(x)(E_{n-k-1}^{*}\phi_i)(x)\right]_{x_i}^{x_{i+1}} \\ &= \sum_{k=0}^{n-1}\left\{-\sum_{i=1}^{m+1} u^{(k)}(x_i)\left[(E_{n-k-1}^{*}\phi_{i-1})(x)\right]_{x=x_i}\right. \\ &\quad \left.+\sum_{i=0}^{m} u^{(k)}(x_i)\left[(E_{n-k-1}^{*}\phi_i)(x)\right]_{x=x_i}\right\} \\ &= \sum_{k=0}^{n-1}\sum_{i=1}^{m}\left[(E_{n-k-1}^{*}(\phi_i - \phi_{i-1}))(x)\right]_{x=x_i} u^{(k)}(x_i),\end{aligned}$$

and equation (6.1.11) becomes

$$\begin{aligned}&\int_a^b g(x)u(x)dx \\ &= \sum_{k=0}^{n-1}\sum_{i=1}^{m}\left\{\left(E_{n-k-1}^{*}(\phi_i - \phi_{i-1})\right)(x)\right\}_{x=x_i} u^{(k)}(x_i) \\ &\quad +\sum_{i=0}^{m}\int_{x_i}^{x_{i+1}} \phi_i(x)(Eu)(x)dx. \qquad (6.1.12)\end{aligned}$$

Denote

$$A_{ki} = \left\{\left(E_{n-k-1}^{*}(\phi_i - \phi_{i-1})\right)(x)\right\}_{x=x_i} \qquad (6.1.13)$$

for $k = 0, 1, \ldots, n-1$ and $i = 1, 2, \ldots, m$, and denote

$$R[u(x)] = \sum_{i=0}^{m}\int_{x_i}^{x_{i+1}} \phi_i(x)(Eu)(x)dx. \qquad (6.1.14)$$

Obviously, A_{ki} is independent of $u(x)$, and $(Eu)(x) = 0$ implies that $R[u(x)] = 0$. We thus have the following rule for constructing an elementary quadrature formula with respect to nodes $x_1, x_2, \ldots, x_m$ and to the linear differential operator E.

(1) By using formulas (6.1.10), we can find $\phi_0(x)$ and $\phi_m(x)$, two solutions of the differential equation $(E^*\phi)(x) = g(x)$ that satisfy initial conditions $\phi_0^{(k)}(a) = 0$ and $\phi_m^{(k)}(b) = 0$ $(k = 0, 1, \ldots, n-1)$;

(2) We arbitrarily fix $m-1$ other solutions $\phi_1(x), \ldots, \phi_{m-1}(x)$ of the same equation;

(3) By using formula (6.1.13), we evaluate all coefficients A_{ki}, for $k = 0, 1, \ldots, n-1$ and $i = 1, 2, \ldots, m$;

(4) Finally, write the remainder from equation (6.1.14).

As was described in [22], the above rule is not only simple but is also the only one possible. The existence and uniqueness will be proved in the following theorem.

Theorem 6.1.1 *Let $x_1, \ldots, x_m$ be m given nodes in $[a, b]$, and A_{ki} $(k = 0, 1, \ldots, n-1$ and $i = 1, 2, \ldots, m)$ be arbitrary nm constants. If the linear functional*

$$R(u) = \int_a^b g(x)u(x)dx - \sum_{k=0}^{n-1}\sum_{i=1}^{m} A_{ki}u^{(k)}(x_i) \tag{6.1.15}$$

vanishes for all solutions of the differential equation $(Eu)(x) = 0$, then there uniquely exist $m-1$ solutions $\phi_1(x), \ldots, \phi_{m-1}(x)$ of the differential equation $(E^*\phi)(x) = g(x)$ that together with the functions $\phi_0(x)$ and $\phi_m(x)$ shown in (6.1.10) make equations (6.1.13) and (6.1.14) hold.

Proof. Equation (6.1.13) can be rewritten as

$$\left(E^*_{n-k-1}\phi_i\right)(x_i) = A_{ki} + \left(E^*_{n-k-1}\phi_{i-1}\right)(x_i),$$

where $k = 0, 1, \ldots, n-1$ and $i = 1, 2, \ldots, m$. We can use the following method to find $\phi_1(x), \ldots, \phi_m(x)$ that satisfy the above equations. Since $\phi_0(x)$ is determined by (6.1.10), $\phi_1(x)$ is the unique solution of the initial problem

$$\begin{cases} \left(E^*\phi_1\right)(x) = g(x), \\ \left(EE^*_{n-k-1}\phi_1\right)(x_1) = A_{k1} + \left(E^*_{n-k-1}\phi_0\right)(x_1) \end{cases}$$

$(k = 0, 1, \ldots, n-1)$. Similarly, we can determine unique $\phi_2(x), \ldots,$ and $\phi_{m-1}(x)$. After we find $\phi_m(x)$, we need to prove that it obeys the conditions in the statement of the theorem. In other words, we need to show that

$$\left(E^*_{n-k-1}\phi_m\right)(b) = 0, \quad k = 0, \ldots, n-1, \tag{6.1.16}$$

and we also have to prove (6.1.14).

Obviously, using the solutions $\phi_1(x), \ldots, \phi_m(x)$ just introduced and the same argument derives again equation (6.1.11). However, since equation (6.1.16) has not yet been proved, we cannot obtain (6.1.12) from (6.1.11); therefore, instead of (6.1.12), we have

$$\begin{aligned}
&\int_a^b g(x)u(x)dx \\
= &\sum_{k=0}^{n-1}\sum_{i=1}^{m}\left(E^*_{n-k-1}(\phi_i - \phi_{i-1})\right)(x_i)u^{(k)}(x_i) \\
&-\sum_{k=0}^{n-1}\left(E^*_{n-k-1}\phi_m\right)(b)u^{(k)}(b) \\
&+\sum_{i=0}^{m}\int_{x_i}^{x_{i+1}} \phi_i(x)(Eu)(x)dx. \qquad (6.1.17)
\end{aligned}$$

Comparing equations (6.1.17) and (6.1.15) and noting that all of the solutions, $\phi_0(x), \ldots, \phi_m(x)$, satisfy equation (6.1.16), we also have

$$\begin{aligned}
R(u) \quad = \quad &-\sum_{k=0}^{n-1}\left(E^*_{n-k-1}\phi_m\right)(b)u^{(k)}(b) \\
&+\sum_{i=0}^{m}\int_{x_i}^{x_{i+1}} \phi_i(x)(Eu)(x)dx. \qquad (6.1.18)
\end{aligned}$$

One of the theorem conditions is that $R(u)$ must be vanishing if u satisfies the equation $(Eu)(x) = 0$. It follows that for all functions in a fundamental system of solutions, $\{u_i(x)\}_{i=1}^n$, of $(Eu)(x) = 0$, equation (6.1.18) implies

$$\sum_{k=0}^{n-1}\left(E^*_{n-k-1}\phi_m\right)(b)u_i^{(k)}(b) = 0,$$

$i = 1, \ldots, n$. Because the Wronskian $det\left(u_i^{(k)}(b)\right)$ is not zero, we conclude that equation (6.1.16) holds and expression (6.1.18) reduces to formula (6.1.14). We have completed the proof of the theorem. □

More details on constructing quadrature formulas of definite integrals by using solutions of ordinary differential equations can be found in Ghizzetti and Ossicini's book [22]. In the rest of this chapter, we will describe the extension of Ghizzetti and Ossicini's results; i.e., construct the DRE for higher dimensional integrals by using the solutions of partial differential equations.

6.2 DREs and partial differential equations

This section covers Kratz's result and its generalization and extension in [80] on the construction of DREs by using the solutions of partial differential equations. Following the notations of Section 1.3, we denote Ω to be a bounded and closed region in $\mathbb{R}^n$. Suppose that the boundary of Ω, $\partial\Omega$, can be described by a system of parametric equations. In particular, the points $(x_1, \ldots, x_n)$ on $\partial\Omega$ satisfy the equation

$$\Phi(x_1, \ldots, x_n) = 0, \tag{6.2.1}$$

where Φ has continuous partial derivatives. In addition, $\Phi(x_1, \ldots, x_n) \leq 0$ for all points in Ω.

We begin with the DRE related to the second order differential operator L, defined by

$$Lu = \sum_{i,j=1}^{n} a_{ij}(X)\frac{\partial^2 u}{\partial x_i \partial x_j} + \sum_{i=1}^{n} b_i(X)\frac{\partial u}{\partial x_i} + c(X)u, \tag{6.2.2}$$

where $a_{ij}(X) \in H_n^2(\Omega)$; $b_i(X)$, $c(X) \in H_n^1(\Omega)$; and $H_n^\alpha(\Omega)$, $(\alpha \geq 1)$, is the collection of all functions $f(X) = f(x_1, \ldots, x_n)$ that have continuous partial derivatives $D^{(i_1,\ldots,i_n)}f$, $0 \leq i_k \leq \alpha$, $k = 1, 2, \ldots, n$.

It is well known that the adjoint operator of L can be defined by

$$Mv = \sum_{i,j=1}^{n} \frac{\partial^2(va_{ij}(X))}{\partial x_i \partial x_j} - \sum_{i=1}^{n} \frac{\partial(vb_i(X))}{\partial x_i} + c(X)v. \tag{6.2.3}$$

If we denote by $r_i(X)$ the expression

$$r_i(X) = -v\sum_{j=1}^{n} a_{ij}\frac{\partial u}{\partial x_j} + u\sum_{j=1}^{n} a_{ij}\frac{\partial v}{\partial x_j} + uv\sum_{j=1}^{n} \frac{\partial a_{ij}}{\partial x_j} - b_i uv, \tag{6.2.4}$$

then we have

$$uMv - vLu = \sum_{i=1}^{n} \frac{\partial r_i}{\partial x_i}. \tag{6.2.5}$$

From this relation and Green's formula, we have the following result.

Theorem 6.2.1 *Let $\Omega \in \mathbb{R}^n$ be an n-dimensional bounded closed domain with the boundary $\partial\Omega$ being a piecewise smooth surface. In particular, if $n = 2$, $\partial\Omega$ is a simple closed curve with finite length. Let $u = u(X)$ and $v = v(X)$ be functions in $C^2(\Omega)$, and let L and M be differential operators defined by (6.2.2) and (6.2.3), respectively. Then we have the identity*

$$\int_\Omega (uMv - vLu)\, dX = \int_{\partial\Omega} \left(\sum_{i=1}^{n} r_i \frac{\partial x_i}{\partial \nu}\right) dS, \tag{6.2.6}$$

where $r_i(X)$ is given by (6.2.4). Using the notation

$$p_i = p_i(X) = \sum_{j=1}^{n} a_{ij}(X)\frac{\partial v}{\partial x_j} + \sum_{j=1}^{n} v\frac{\partial a_{ij}(X)}{\partial x_j} - b_i(X)v, \tag{6.2.7}$$

we can rewrite identity (6.2.6) as

$$\int_{\Omega}(uMv - vLu)dX = \int_{\partial\Omega}\left[u\sum_{i=1}^{n} p_i\frac{\partial x_i}{\partial \nu}\right] dS - \int_{\partial\Omega}\left[v\sum_{i=1}^{n}\left(\sum_{j=1}^{n} a_{ij}(X)\frac{\partial u}{\partial x_j}\right)\frac{\partial x_i}{\partial \nu}\right] dS. \tag{6.2.8}$$

Furthermore, if $v = v(X)$ satisfies $Mv = 1$ on Ω, then for any solution, $u = u(X)$, $X \in \Omega$, of $Lu = g$, the following identity holds.

$$\begin{aligned}\int_{\Omega} u(X)\, dV &= \int_{\Omega} v(X)g(X)\, dV + \int_{\partial\Omega}\left[u(X)\sum_{i=1}^{n} p_i(X)\frac{\partial x_i}{\partial \nu}\right] dS \\ &\quad - \int_{\partial\Omega}\left[v(X)\sum_{i=1}^{n}\left(\sum_{j=1}^{n} a_{ij}(X)\frac{\partial u}{\partial x_j}\right)\frac{\partial x_i}{\partial \nu}\right] dS. \end{aligned}\tag{6.2.9}$$

Proof. Consider Green's formula

$$\int_{\Omega}\frac{\partial f(X)}{\partial x_i}\, dX = \int_{\partial\Omega} f(X)\frac{\partial x_i}{\partial \nu}\, dS,$$

where $\frac{\partial x_i}{\partial \nu}$, the outer normal derivative of x_i on the surface $\partial\Omega$, has the expression

$$\frac{\partial x_i}{\partial \nu} = \frac{\partial \phi}{\partial x_i}\left[\left(\frac{\partial \phi}{\partial x_1}\right)^2 + \cdots + \left(\frac{\partial \phi}{\partial x_n}\right)^2\right]^{-1/2}.$$

Replacing $f(X)$ in the formula with $r_i(X)$ gives (6.2.6). Noting the expression of $r_i(X)$ given by (6.2.4), we also obtain identity (6.2.8). In particular, if $v = v(X)$ is a solution to the differential equation $Mv = 1$, then identity (6.2.8) can be reduced to (6.2.9), which is obviously a DRE with the remainder $\int_{\Omega} v(X)g(X)\, dX$. □

If the remainder term of identity (6.2.9) vanishes (i.e., $g(X)$ satisfies $\int_{\Omega} v(X)g(X)dV = 0$), we obtain the following exact DRE for the integral $\int_{\Omega} u\, dX$.

$$\begin{aligned}\int_{\Omega} u\, dX &= \int_{\partial\Omega}\left[u\sum_{i=1}^{n} p_i\frac{\partial x_i}{\partial \nu}\right] dS \\ &\quad - \int_{\partial\Omega}\left[v\sum_{i=1}^{n}\left(\sum_{j=1}^{n} a_{ij}(X)\frac{\partial u}{\partial x_j}\right)\frac{\partial x_i}{\partial \nu}\right] dS. \end{aligned}\tag{6.2.10}$$

In addition, if v satisfies the boundary condition $v(X) = 0$, $X \in \partial\Omega$ (i.e., v is the solution of the boundary value problem $Mv = 1$ and $v = 0$ on $\partial\Omega$), then expansion (6.2.10) can be reduced to

$$\int_\Omega u\, dX = \int_{\partial\Omega} \left[u \sum_{i=1}^n \left(\sum_{j=1}^n a_{ij}(X) \frac{\partial v}{\partial x_j} \right) \frac{\partial x_i}{\partial \nu} \right] dS. \tag{6.2.11}$$

The above exact DRE is convenient for computing $\int_\Omega u(X) dV$.

In the following, we will specialize formula (6.2.9) by choosing L to be classic partial differential operators such as elliptic, hyperbolic, and parabolic operators.

Corollary 6.2.2 *Let L, as shown in (6.2.2), be an* elliptic operator, *in which*

$$a_{ij} = \begin{cases} 1, & \text{if } i = j, \\ 0, & \text{if } i \neq j, \end{cases}$$

and let v be a solution of the Dirichlet problem

$$\begin{cases} \sum_{i=1}^n \dfrac{\partial^2 v}{\partial x_i^2} - \sum_{i=1}^n \dfrac{\partial(v b_i)}{\partial x_i} + cv = 1, & \text{on } \Omega; \\ v = 0, & \text{on } \partial\Omega. \end{cases} \tag{6.2.12}$$

Then,

$$\int_\Omega u(X) dV = \int_\Omega v(X) g(X) dV + \int_{\partial\Omega} \left[u(X) \sum_{i=1}^n \left(\frac{\partial v}{\partial x_i} \frac{\partial x_i}{\partial \nu} \right) \right] dS, \tag{6.2.13}$$

where the function $u = u(X)$ is a solution of the equation

$$\sum_{i=1}^n \frac{\partial^2 u}{\partial x_i^2} + \sum_{i=1}^n b_i \frac{\partial u}{\partial x_i} + cu = g.$$

In the two-dimensional case, the corresponding elliptic operator is

$$Lu = u_{xx} + u_{yy} + Du_x + Eu_y + Fu,$$

and formula (6.2.13) becomes

$$\iint_S u(x,y) dx dy = \iint_S v(x,y) g(x,y) dx dy + \int_T u(x,y) \left(-v_y(x,y) dx + v_x(x,y) dy \right). \tag{6.2.14}$$

Here $v = v(x, y)$ is the solution of the Dirichlet problem

$$\begin{cases} v_{xx} + v_{yy} - (Dv)_x - (Ev)_y + Fv = 1, & \text{on } S, \\ v = 0, & \text{on } T, \end{cases} \tag{6.2.15}$$

and $u = u(x, y)$ is any solution of the equation $Lu = u_{xx} + u_{yy} + Du_x + Eu_y + Fu = g$ on S. Throughout the rest of this section, S is assumed to be the closure of a simply-connected bounded open set in the plane with the boundary T.

Example 6.2.1 (The *reduced wave equation* on a disc.) Let $Lu = u_{xx}+u_{yy}+\lambda^2 u$ on $S = \{(x, y) : x^2+y^2 \le R^2\}$. Here, we assume that $\lambda \neq 0$ and λR is not a zero of the Bessel function J_0. If we use the polar coordinate system, then problem (6.2.15) can be written as

$$\begin{cases} v_{rr} + r^{-1}v_r + r^{-2}v_{\theta\theta} + \lambda^2 v = 1, & \text{on } S, \\ v = 0, & \text{on } T. \end{cases}$$

This problem then has the solution

$$v(r, \theta) = \lambda^{-2}[1 - J_0(\lambda r)]/J_0(\lambda R).$$

For all solutions of $Lu = 0$ on S, formula (6.2.14) gives the exact DRE,

$$\iint_S u(x, y)dxdy = -\frac{RJ_0'(\lambda R)}{\lambda J_0(\lambda R)} \int_0^{2\pi} u(R, \theta)d\theta.$$

Corollary 6.2.3 *Let L in (6.2.2) be a* parabolic operator, *in which*

$$a_{ij} = \begin{cases} 1, & \textit{if } 1 \le i = j \le n-m,\ 0 < m < n, \\ 0, & \textit{otherwise,} \end{cases}$$

and let v be a solution to the problem

$$\begin{cases} \displaystyle\sum_{i=1}^{n-m} \frac{\partial^2 v}{\partial x_i^2} - \sum_{i=1}^{n} \frac{\partial(vb_i)}{\partial x_i} + cv = 1, \quad \textit{if } X \in \Omega, \\ v(0, x_2, \dots, x_n) = \cdots = v(x_1, \dots, x_{n-m-1}, 0, x_{n-m+1}, \dots, x_n) \\ = v(1, x_2, \dots, x_n) = \cdots = v(x_1, \dots, x_{n-m-1}, 1, x_{n-m+1}, \cdots, x_n) = 1, \end{cases} \tag{6.2.16}$$

where $\Omega = \{X = (x_1, \dots, x_n) : 0 \le x_i \le 1, i = 1, \dots, n\}$. *Then for all* $u = u(X)$ *that satisfy*

$$\sum_{i=1}^{n-m} \frac{\partial^u}{\partial x_i^2} + \sum_{i=1}^{n} b_i \frac{\partial u}{\partial x_i} + cu = g,$$

we have

$$\int_\Omega u(X)\,dV = \int_\Omega v(X)g(X)dV + \int_{\partial\Omega} \left[u \sum_{i=1}^{n-m} \left(\frac{\partial v}{\partial x_i} \frac{\partial x_i}{\partial \nu} \right) - \sum_{i=n-m+1}^{n} \left(b_i v \frac{\partial x_i}{\partial \nu} \right) \right] dS. \tag{6.2.17}$$

For the two-dimensional case, the corresponding parabolic operator becomes

$$Lu = u_{xx} + Du_x + Eu_y + Fu,$$

and DRE formula (6.2.17) can be reduced to

$$\iint_S u dx dy = \iint_S v g dx dy + \int_T u(E v dx + v_x dy). \tag{6.2.18}$$

$u = u(x, y)$ is any solution of $Lu = g$ on S, and $v = v(x, y)$ satisfies

$$\begin{cases} v_{xx} - (Dv)_x - (Ev)_y + Fv = 1, & \text{on } S = [a, b] \times [c, d], \\ v(a, y) = v(b, y) = 0, & \text{for all } y \in [c, d]. \end{cases}$$

Example 6.2.2 (The *heat equation* on a rectangle.) Let $Lu = u_{xx} - \lambda u_y$ on $S = [a, b] \times [c, d]$. The function $v = v(x, y) = (x - a)(x - b)/2$ satisfies $Mv = v_{xx} + \lambda v_y = 1$ and vanishes on the vertical sides of S. Formula (6.2.18) now generates the exact DRE

$$\begin{aligned} \iint_S u dx dy &= \tfrac{\lambda}{2} \textstyle\int_a^b (x - a)(x - b) \left[u(x, d) - u(x, c)\right] dx \\ &\quad + \frac{b - a}{2} \int_c^d \left[u(b, y) - u(a, y)\right] dy \end{aligned}$$

for all solutions of $Lu = 0$ on S.

Corollary 6.2.4 *Let L in (6.2.2) be a* hyperbolic operator, *in which*

$$a_{ij} = \begin{cases} 1, & \text{if } i = j = 1, \\ -1, & \text{if } i = j \neq 1, \\ 0, & \text{otherwise,} \end{cases}$$

and let v be a solution of the equation

$$\frac{\partial^2 v}{\partial x_1^2} - \sum_{i=2}^{n} \frac{\partial^2 v}{\partial x_i^2} - \sum_{i=1}^{n} \frac{\partial (v b_i)}{\partial x_i} + cv = 1, \quad X \in \Omega. \tag{6.2.19}$$

Then

$$\begin{aligned} \int_\Omega u(X) dV &= \int_\Omega v(X) g(X) dV \\ &\quad + \int_{\partial\Omega} u \left[\left(\frac{\partial v}{\partial x_1} - b_1 v \right) \frac{\partial x_1}{\partial \nu} - \sum_{i=2}^{n} \left(\frac{\partial v}{\partial x_i} + b_i v \right) \frac{\partial x_i}{\partial \nu} \right] dS \\ &\quad - \int_{\partial\Omega} v \left[\frac{\partial u}{\partial x_1} \frac{\partial x_1}{\partial \nu} - \sum_{i=2}^{n} \frac{\partial u}{\partial x_i} \frac{\partial x_i}{\partial \nu} \right] dS \end{aligned} \tag{6.2.20}$$

holds for any solution $u = u(X)$ *of*

$$\frac{\partial^2 u}{\partial x_1^2} - \sum_{i=2}^{n} \frac{\partial^2 u}{\partial x_i^2} + \sum_{i=1}^{n} b_i \frac{\partial u}{\partial x_i} + cu = g.$$

In the two-dimensional case, the hyperbolic operator is reduced to

$$Lu = u_{xy} + Du_x + Eu_y + Fu.$$

Replacing Ω in DRE (6.2.20) by $S = [a, b] \times [c, d]$ yields

$$\begin{aligned} \iint_S u dx dy &= -[uv]_{x=a,y=c}^{x=b,y=d} + \iint_S vg dx dy \\ &\quad + \int_T u\left[(Ev - v_x)dx + ((v_y - Dv)dy\right], \qquad (6.2.21) \end{aligned}$$

where $v = v(x, y)$ satisfies

$$v_{xy} - (Dv)_x - (Ev)_y + Fv = 1$$

on S; and $u = u(x, y)$ is a solution of $Lu = g$ on S.

Example 6.2.3 (The *Darboux equation* on a rectangle.) Let $Lu = u_{xy} + \lambda(x + y)^{-1}(u_x + u_y)$, where $\lambda > 1$; and, to keep the analysis simple, assume that the rectangle $S = [a, b] \times [c, d]$ does not intersect the line $x + y = 0$. Clearly, $v = v(x, y) = (x + y)^2/(2 - 2\lambda)$ is a solution of the adjoint equation

$$Mv = v_{xy} - \lambda(x + y)^{-1}(v_x + v_y) + 2\lambda(x + y)^{-2}v = 1.$$

Therefore, formula (6.2.20) generates the exact DRE

$$\begin{aligned} &\iint_S u(x, y) dx dy \\ &= \frac{2 - \lambda}{2 - 2\lambda}\left[\int_a^b (x + d)u(x, d)dx - \int_a^b (x + c)u(x, c)dx \right. \\ &\quad \left. + \int_c^d (b + y)u(b, y)dy - \int_c^d (a + y)u(a, y)dy\right] \\ &\quad + \frac{1}{2\lambda - 2}\left[(b + d)^2 u(b, d) - (b + c)^2 u(b, c) \right. \\ &\quad \left. - (a + d)^2 u(a, d) + (a + c)^2 u(a, c)\right] \end{aligned}$$

for all solutions of $Lu = 0$ on S.

Finally, let us consider an example of a first-order partial differential equation.

Corollary 6.2.5 *If* $v = v(X)$ *is a solution of the equation*

$$-\sum_{i=1}^{n} \frac{\partial b_i v}{\partial x_i} + cv = 1, \quad X \in \Omega,$$

then for any function $u = u(X)$ *that satisfies*

$$\sum_{i=1}^{n} b_i(X) \frac{\partial u}{\partial x_i} + cu = 0$$

we have

$$\int_{\Omega} u dV = -\int_{\partial\Omega} \left(uv \sum_{i=1}^{n} b_i \frac{\partial x_i}{\partial \nu} \right) dS.$$

For the two-dimensional case, Corollary 6.2.5 tells us that if a bivariate function $v = v(x, y)$ satisfies

$$-(Dv)_x - (Ev)_y + Fv = 1 \tag{6.2.22}$$

on S, then for all solutions $u = u(x, y)$ of the equation

$$Du_x + Eu_y + Fu = 0 \tag{6.2.23}$$

we have the exact DRE

$$\iint_S u dx dy = \int_T uv(E dx - D dy). \tag{6.2.24}$$

We have established a DRE that is valid for the integrals of all solutions of $Lu = g$ with pre-specified operator L and function g. If $\langle g, v \rangle = 0$, then the DRE is exact. Here, v satisfies the adjoint equation $Mv = 1$. We now turn to a more practical problem: can we develop such a DRE for an arbitrary function u when L and g are not given? The answer is affirmative provided that we can come up with an operator L, tailored to u, such that the adjoint equation $Mv = 1$ can be solved. We now give the details for the two-dimensional case. Influenced by Corollary 6.2.5, an appropriate choice of L for a differentiable function is the operator $L = u_y(\partial/\partial x) - u_x(\partial/\partial y)$. For then $Lu = 0$, and if v satisfies

$$Mv = -u_y v_x + u_x v_y = 1,$$

DRE (6.2.24) yields

$$\iint_S u dx dy = \int_T uv(-u_x dx - u_y dy).$$

Let S be a rectangle situated parallel to the axes. An integration by parts yields

$$\iint_S u dx dy = \frac{1}{2} \int_T u^2 (v_x dx + v_y dy). \tag{6.2.25}$$

We now extend DRE (6.2.25) to the following result.

Theorem 6.2.6 *Let S be the closure of a simply-connected bounded open set in the plane with boundary T. Let $u \in C^1(S)$ and ϕ be a differentiable univariate function. Suppose $v \in C^2(S)$ satisfies*

$$\left(-u_y v_x + u_x v_y\right) \phi'(u) = u. \tag{6.2.26}$$

Then

$$\iint_S u dx dy = \int_T \phi(u) dv. \tag{6.2.27}$$

Proof. Applying Green's theorem yields

$$\begin{aligned}
&\int_T \phi(u) dv \\
= &\int_T \left[\phi(u) v_x dx + \phi(u) v_y dy\right] \\
= &\iint_S \left[\phi'(u) u_x v_y + \phi(u) v_{yx} - \phi'(u) u_y v_x - \phi(u) v_{xy}\right] dx dy \\
= &\iint_S u dx dy.
\end{aligned}$$

The theorem is thereby proved. □

We should point out that it is sometimes very hard or even impossible to find a v satisfying equation (6.2.26). For instance, such a v does not exist for u that satisfies $grad\, u = 0$ at a point in S at which $u \neq 0$. Hence, the utility of Theorem 6.2.6 is limited. If v cannot be found analytically, one can attempt an approximation of v by numerical methods, but this may entail greater expanse than the direct numerical approximation of $\int\int_S u dx dy$.

There are, however, numerous integrands that respond favorably to Theorem 6.2.6 as illustrated by the following examples.

Example 6.2.4 Let $u = \lambda x + \mu y$, where λ and μ are constants. Setting $\phi_1(t) = t^2/1$ and $\phi_2(t) = t$, we solve equation (6.2.26) to get

$$v_1 = (\mu x - \lambda y)/(\lambda^2 + \mu^2)$$

and

$$v_2 = [-\lambda\mu x^2 + (\lambda^2 - \mu^2) x y + \lambda\mu y^2]/(\lambda^2 + \mu^2).$$

Thus, DRE (6.2.27) gives

$$\begin{aligned}
\iint_S u dx dy &= \tfrac{1}{2} \int_T u^2 dv_1 \\
&= \int_T u dv_2.
\end{aligned} \tag{6.2.28}$$

Example 6.2.5 Let $u = x^2 + y^2$, and assume $(0, 0) \notin S$. From equation (6.2.26) with $\phi(t) = t$, we find a solution

$$v(x, y) = -\frac{1}{2}(x^2 + y^2)a(x, y), \tag{6.2.29}$$

where $a(x, y)$ is a continuous determination of the inverse tangent of x/y. We obtain this DRE for the function u: $\int\int_S u dx dy = \int_T u dv$.

Before giving more examples, we establish the following lemma, which can be proved by direct calculation. The lemma will help us to find the solution v in many instances and to establish an exact DRE using formula (6.2.27).

Lemma 6.2.7 *Let θ and ψ be univariate functions, and let u, v, w be bivariate functions for which det $[\partial(u, w)/\partial(x, y)] = \theta(u)$ and*

$$v = \frac{\psi(u)w}{\psi'(u)\theta(u)}.$$

Then det $[\partial(\psi(u), v)/\partial(x, y)] = \psi(u)$.

Example 6.2.6 We will apply DRE (6.2.27) to integral $\int\int_S \exp(-x^2 - y^2) dx dy$, where $(0, 0) \notin S$. Equations (6.2.28) and (6.2.29) imply

$$\left| \frac{\partial \left(x^2 + y^2, -\frac{1}{2}(x^2 + y^2)a(x, y)\right)}{\partial(x, y)} \right| = x^2 + y^2.$$

It follows from Lemma 6.2.7, with $\theta(t) = t$ and $\psi(t) = \exp(-t)$, that

$$\left| \frac{\partial \left(\exp(-x^2 - y^2), v\right)}{\partial(x, y)} \right| = \exp(-x^2 - y^2),$$

where $v = \frac{1}{2}a(x, y)$. With this choice of v, DRE formula (6.2.27) yields

$$\iint_S \exp(-x^2 - y^2) dx dy = \int_T \exp(-x^2 - y^2) dv. \tag{6.2.30}$$

We need to point out that the restriction $(0, 0) \notin S$ can be avoided by removing a narrow corridor R containing $(0, 0)$ and connecting $(0, 0)$ to T; the error so introduced is controlled by

$$\left| \iint_S u dx dy - \iint_{S \backslash R} u dx dy \right| \leq \max_R |u| \cdot (\text{area of } R). \tag{6.2.31}$$

We will compare on several different regions S the exact DRE formula with the standard techniques for evaluating the integral $\iint_S u dx dy$.

We first use the trapezoidal approximation on the Stieltjes integral $\int_T u dv$. Parameterizing T by t on $[0, 1]$ with nodes $t_i = i/N$ $(i = 0, 1, \ldots, N)$ and defining $u_i = u(x(t_i), y(t_i))$ and $v_i = v(x(t_i), y(t_i))$, we set

$$K_N \equiv \sum_{i=0}^{N-1} (v_{i+1} - v_i)(u_{i+1} + u_i)/2 \approx \int_T u dv. \tag{6.2.32}$$

Thus, from (6.2.30) and (6.2.32) we have quadrature formula

$$\iint_S \exp(-x^2 - y^2) dx dy \approx K_N, \tag{6.2.33}$$

where K_N is given by equation (6.2.32) with

$$u = \exp(-x^2 - y^2), \quad v = \frac{1}{2} a(x, y). \tag{6.2.34}$$

We remark here that a much better numerical technique, such as an extrapolation process or a Gaussian quadrature, can be applied to integral $\int_T u dv$. However, we prefer to use unsophisticated quadrature techniques here in order to make the comparisons clearer.

Our first example is for $S = [1, 3] \times [1, 2]$. Since the integral

$$\int_1^2 \int_1^3 \exp(-x^2 - y^2) dx dy$$

readily factors into two one-dimensional integrals, the method shown by (6.2.33) is barely competitive with standard techniques. For instance, we have $K_{1,000} = 0.188525$, which is correct to six decimal places. If the trapezoidal rule is applied to evaluate $\int_1^3 \exp(-x^2) dx$ and $\int_1^2 \exp(-y^2) dy$, then about the same number of evaluations are required to achieve the same accuracy.

The second example is for $S = \{(x, y) : x^2 + y^2 \leq 1\}$. There is an extra complication in this problem in that S contains the origin, at which point v is undefined. This difficulty is circumvented by removing from S the slit $R = [-1, 10^{-6}] \times [-10^{-6}, 10^{-6}]$ and integrating over the incised region $S \setminus R$. Thus, an error dominated by $2 \cdot 10^{-6}$ is introduced by estimate (6.2.31). Parameterizing the boundary of $S \setminus R$ in a straightforward fashion, we obtain $K_8 = 0.172523$, $K_{40} = 0.197425$, $K_{200} = 0.198539$, $K_{1,000} = 0.198584$, etc. Note that the correct value to six significant figures is 0.198586.

6.3 Applications of DREs in the construction of BTQFs

First, using DREs (6.2.10) and (6.2.11), we can construct a type of boundary quadrature formula. For instance, let us consider the cubic domain $\Omega = V_n(0 \leq$

$x_i \leq 1, i = 1, \ldots, n$). From equation (6.2.11), we obtain

$$\int_{V_n} u dV = \sum_{i=1}^{n} \int_{V_{n-1}} [F_i(X)]_{x_i=0}^{x_i=1} dV_{n-1}, \tag{6.3.1}$$

where

$$F_i(X) = u(X) \sum_{j=1}^{n} a_{ij}(X) \frac{\partial v}{\partial x_j}(X) \tag{6.3.2}$$

and

$$\begin{aligned} [F_i(X)]_{x_i=0}^{x_i=1} &= F_i(x_1, \ldots, x_{i-1}, 1, x_{i+1}, \ldots, x_n) \\ &\quad - F_i(x_1, \ldots, x_{i-1}, 0, x_{i+1}, \ldots, x_n). \end{aligned}$$

Obviously, $F_i \in H_{n-1}^1(V_{n-1}, C)$. Here, $H_n^\alpha(V_n, C)$ ($\alpha \geq 1$) is the class of all functions $f(x_1, \ldots, x_n)$ that possess the continuous mixed partial derivatives

$$f^{(i_1, \ldots, i_n)} = \frac{\partial^{i_1 + \cdots + i_n}}{\partial x_1^{i_1} \cdots \partial x_n^{i_n}} \quad (0 \leq i_\nu \leq \alpha,\ 1 \leq \nu \leq n)$$

on V_n and $|f^{(i_1, \ldots, i_n)}| \leq C$.

We will use a Halton sequence to construct a quadrature formula of integral (6.3.1). This construction process is a number theory method, and a Halton sequence is a uniformly distributed sequence. Let $\{M_k = (x_{k,1}, \ldots, x_{k,n}), k = 1, 2, \ldots\}$ be a sequence of points in the s-dimensional cubic domain V_s. For any element $(a_1, \ldots, a_n) \in V_n$, we denote by $N_n(a_1, \ldots, a_n)$ the number of elements in $\{M_1, \ldots, M_m\}$ that satisfy the inequalities

$$x_{k,1} < a_1, \ldots, x_{k,n} < a_n. \tag{6.3.3}$$

If

$$\lim_{n \to \infty} \frac{N_m(a_1, \ldots, a_n)}{m} = a_1 \cdots a_n \tag{6.3.4}$$

holds, then we say that $M_k (k = 1, 2, \ldots)$ is uniformly distributed in V_n.

Roughly speaking, a *sequence of uniformly distributed points* is a sequence of points with the same probability distribution everywhere.

If we use a sequence of points as the sequence of nodes in a quadrature formula, then the uniform distribution property of the point sequence determines the error bound of the quadrature formula. We describe this in the following lemma.

Lemma 6.3.1 *Let* $\{M_k, k = 1, 2, \ldots\}$ *be a sequence of points in* V_n. *If*

$$\sup_{0 \leq x_i \leq 1} \left| \frac{N_m(a_1, \ldots, a_n)}{m} - x_1 \cdots x_n \right| < \phi(m), \tag{6.3.5}$$

then

$$\sup_{f\in H_n^1(C)} \left| \int_{V_n} f(X)dX - \frac{1}{m}\sum_{k=1}^{m} f(_{k,1},\ldots,x_{k,n}) \right| \le 2^n C\phi(m). \tag{6.3.6}$$

Proof. We only give a proof for the case of $n = 2$ because the proof for $n > 2$ is similar. We write

$$\begin{aligned} f(x_1,x_2) &= f(1,1) - (f(1,1) - f(x_1,1)) - (f(1,1) - f(1,x_2)) \\ &\quad + (f(x_1,x_2) - f(x_1,1) - f(1,x_2) + f(1,1)) \\ &= f(1,1) - \int_{x_1}^1 f'_{y_1}(y_1,1)dy_1 - \int_{x_2}^1 f'_{y_2}(1,y_2)dy_2 \\ &\quad + \int_{x_1}^1\int_{x_2}^1 f''_{y_1y_2}(y_1,y_2)dy_1dy_2. \end{aligned} \tag{6.3.7}$$

Because

$$\begin{aligned} &\int_0^1\int_0^1\int_{x_1}^1 f'_{y_1}(y_1,1)dy_1dx_1dx_2 \\ &= \int_0^1\int_0^{y_1} f'_{y_1}(y_1,1)dx_1dy_1 \\ &= \int_0^1 y_1 f'_{y_1}(y_1,1)dy_1, \\ &\int_0^1\int_0^1\int_{x_2}^1 f'_{y_2}(1,y_2)dy_2dx_1dx_2 = \int_0^1 y_2 f'_{y_2}(1,y_2)dy_2, \end{aligned}$$

and

$$\begin{aligned} &\int_0^1\int_0^1\int_{x_1}^1\int_{x_2}^1 f''_{y_1y_2}(y_1,y_2)dy_1dy_2dx_1dx_2 \\ &= \int_0^1\int_0^1 y_1y_2 f''_{y_1y_2}(y_1,y_2)dy_1dy_2, \end{aligned}$$

from (6.3.7) we obtain

$$\begin{aligned} &\int_0^1\int_0^1 f(x_1,x_2)dx_1dx_2 \\ = \;& f(1,1) - \int_0^1 y_1 f'_{y_1}(y_1,1)dy_1 - \int_0^1 y_2 f'_{y_2}(1,y_2)dy_2 \\ &+ \int_0^1\int_0^1 y_1y_2 f''_{y_1y_2}(y_1,y_2)dy_1dy_2. \end{aligned} \tag{6.3.8}$$

Using the notation

$$K(u) = \begin{cases} 1, & \text{if } u > 0, \\ 0, & \text{if } u \le 0 \end{cases}$$

changes (6.3.7) to

$$\begin{aligned}
&\frac{1}{m}\sum_{k=1}^{m} f(x_{k,1}, x_{k,2}) \\
&= f(1,1) - \frac{1}{m}\sum_{k=1}^{m}\int_{x_{k,1}}^{1} f'_{y_1}(y_1, 1)dy_1 - \frac{1}{m}\sum_{k=1}^{m}\int_{x_{k,2}}^{1} f'_{y_2}(1, y_2)dy_2 \\
&\quad + \frac{1}{m}\sum_{k=1}^{m}\int_{x_{k,1}}^{1}\int_{x_{k,2}}^{1} f''_{y_1 y_2}(y_1, y_2)dy_1 dy_2 \\
&= f(1,1) - \frac{1}{m}\sum_{k=1}^{m}\int_{0}^{1} K(y_1 - x_{k,1}) f'_{y_1}(y_1, 1)dy_1 \\
&\quad - \frac{1}{m}\sum_{k=1}^{m}\int_{0}^{1} K(y_2 - x_{k,2}) f'_{y_2}(1, y_2)dy_2 \\
&\quad + \frac{1}{m}\sum_{k=1}^{m}\int_{0}^{1}\int_{0}^{1} K(y_1 - x_{k,1})K(y_2 - x_{k,2}) f''_{y_1 y_2}(y_1, y_2)dy_1 dy_2 \\
&= f(1,1) - \frac{1}{m}\int_{0}^{1} N_m(y_1, 1) f'_{y_1}(y_1, 1)dy_1 \\
&\quad - \frac{1}{m}\int_{0}^{1} N_m(1, y_2) f'_{y_2}(1, y_2)dy_2 \\
&\quad + \frac{1}{m}\int_{0}^{1}\int_{0}^{1} N_m(y_1, y_2) f''_{y_1 y_2}(y_1, y_2)dy_1 dy_2.
\end{aligned} \tag{6.3.9}$$

Combining (6.3.8) and (6.3.9),

$$\begin{aligned}
&\left|\int_0^1\int_0^1 f(x_1, x_2)dx_1 dx_2 - \frac{1}{m}\sum_{k=1}^{m} f(x_{k,1}, x_{k,2})\right| \\
&\le \int_0^1 \left|\frac{N_m(y_1, 1)}{m} - y_1\right| \left|f'_{y_1}(y_1, 1)\right| dy_1 \\
&\quad + \int_0^1 \left|\frac{N_m(1, y_2)}{m} - y_2\right| \left|f'_{y_2}(1, y_2)\right| dy_2 \\
&\quad + \int_0^1\int_0^1 \left|\frac{N_m(y_1, y_2)}{m} - y_1 y_2\right| \left|f''_{y_1 y_2}(y_1, y_2)\right| dy_1 dy_2 < 4C\phi(m).
\end{aligned}$$

The proof is complete. □

The $\phi(m)$ shown in Lemma 6.3.1 can be considered as a measurement of the "degree"of the uniform distribution of a point sequence. Therefore, the construction of numerical quadrature formulas is equivalent to seeking a sequence of points with a high "degree" of uniform distribution. Before finding such a sequence, we need to know what is the highest possible approximation order of a numerical quadrature with the nodes $\{M_k\}_{k=1}^N$. The following lemma shows that the highest possible order is usually $O(N^{-1})$ even when the integrand is infinitely differentiable.

Lemma 6.3.2 *Let $M_k = (x_{k,1}, \dots, x_{k,n})$ $(k = 1, 2, \dots)$ be an arbitrary sequence of points. Then, for the class of analytic functions, the order of the error term ρ_N of the numerical quadrature formula*

$$\int_{V_n} f(X)dX = \frac{1}{N}\sum_{k=1}^{N} f(x_{k,1}, \dots, x_{k,n}) + \rho_N \tag{6.3.10}$$

cannot be higher than $O(N^{-1})$.

Proof. If the conclusion of the lemma is not correct, we can have a point sequence $M_k = (x_{k,1}, \dots, x_{k,n})$ $(k = 1, 2, \dots)$ such that

$$\int_{V_n} f(X)dX = \frac{1}{N}\sum_{k=1}^{N} f(x_{k,1}, \dots, x_{k,n}) + \frac{1}{N}. \tag{6.3.11}$$

Multiplying both sides of equation (6.3.11) by N and replacing $f(x_1, \dots, x_n)$ by $\phi(x_1)$, we obtain

$$\sum_{k=1}^{N} \phi(x_{k,1}) = N \int_0^1 \phi(x_1)dx_1 + o(1).$$

It follows that

$$\begin{aligned} \phi(x_{N,1}) &= \sum_{k=1}^{N} \phi(x_{k,1}) - \sum_{k=1}^{N-1} \phi(x_{k,1}) \\ &= \int_0^1 \phi(x_1)dx_1 + o(1). \end{aligned}$$

In the above equation, replacing $\phi(x_1)$ by $\sin(2\pi x_1)$ gives

$$\sin(2\pi x_{N,1}) = o(1). \tag{6.3.12}$$

Similarly, substituting in $\phi(x_1) = \cos 2\pi x_1$ yields

$$\cos(2\pi x_{N,1}) = o(1). \tag{6.3.13}$$

Then, from equations (6.3.12) and (6.3.13),

$$1 = \sin^2(2\pi x_{N,1}) + \cos^2(2\pi x_{N,1}) = o(1).$$

This contradiction shows that the order of the error ρ_N cannot be higher than $O(N^{-1})$. □

A point sequence is called the optimal uniformly distributed point sequence if it possesses the highest possible degree of uniform distribution; i.e., it yields the numerical quadrature formula with the best approximation order (see Lemma 6.3.1). We now introduce *Halton point sequences*, which are optimal uniformly distributed point sequences. They are constructed by representing the numbers in base r.

Let $r > 1$. Then any natural number $k \leq m$ can be represented in base r as

$$k = k_0 + k_1 r + \cdots + k_M r^M, \quad 0 \leq k_j < r, 0 \leq j \leq M, \tag{6.3.14}$$

where $M = [\log_r m]$. Define

$$\phi_r(k) = k_0 r^{-1} + k_1 r^{-2} + \cdots + k_M r^{-M-1}. \tag{6.3.15}$$

Then every whole number corresponds to a unique decimal in base r; i.e., $k \leftrightarrow \phi_r(k)$. Here $\phi_r(k)$ can be considered as the image of k. For the first n prime numbers $p_1, \ldots, p_n$, we call the sequence $\{M_k = (\phi_{p_1}(k), \ldots, \phi_{p_n}(k)), k = 1, 2, \ldots\}$ a Halton sequence. We now check the "degree" of the uniform distribution of the Halton sequences.

Theorem 6.3.3 *Let p_i be the ith prime number. Then, if $m > p_n$, for the Halton sequence $M_k = (\phi_{p_1}(k), \ldots, \phi_{p_n}(k))$ $(k = 1, 2, \cdots)$, we have*

$$\sup_{0 \leq x_j \leq 1} \left| \frac{N_m(x_1, \ldots, x_n)}{m} - x_1 \cdots x_n \right| \leq 2^n \left(\Pi_{i=1}^n \frac{p_i}{\log p_i} \right) \frac{\log^n m}{m}.$$

Proof. Choose an arbitrary infinite decimal in base r,

$$x = 0.a_0 a_1 \cdots a_M \cdots.$$

From (6.3.14) and (6.3.15), if $x > \phi_r(k)$, then one of the following conditions must hold.

$$\begin{cases} a_0 > k_0; \quad a_0 = k_0,\ a_1 > k_1; \ldots; \\ a_0 = k_0, \ldots, a_{M-1} = k_{M-1},\ a_M > k_M; \\ a_0 = k_0, \ldots, a_M = k_M. \end{cases} \tag{6.3.16}$$

In other words, for an integer ℓ with $1 \leq \ell < M + 3$, k possesses the form

$$k \equiv a_0 + \cdots + a_{\ell-2} r^{\ell-2} + k_{\ell-1} r^{\ell-1} (mod\, r^\ell), \tag{6.3.17}$$

where

$$a_{\ell-1} > k_{\ell-1}, \quad k_{M+1} = 0.$$

On the other hand, for given x and k, only one condition in (6.3.16) (i.e., (6.3.17)) can be satisfied. Moreover, for any given ℓ, equation (6.3.17) shows that k belongs to one of the $a_{\ell-1}$ residue classes (or congruence classes) $mod\ r^{\ell}$. Let q be the representative of a residue class. Then in the set $\{1, 2, \ldots, m\}$ there are $\left[\frac{m}{r^{\ell}}\right] + \theta$ ($\theta = 0$ or 1) numbers that are congruent $mod\ r^{\ell}$ with q.

We now count the number of integers, k, in the set $\{1, 2, \ldots, m\}$ that satisfy

$$x_i > \phi_{p_i}(k) \quad (1 \le i \le n),$$

where all x_i $(1 \le i \le n)$ are supposed to be infinite decimals. Here, we need to apply the Chinese remainder theorem. This theorem states that if the greatest common divisor of ℓ_i and ℓ_j, $(\ell_i, \ell_j) = 1$ for $i \neq j$, then the congruence system $x \equiv a_i (mod\ \ell_i)$ $(1 \le i \le m)$ has the unique solution $mod\ \ell_1 \cdots \ell_m$. Thus, the congruence system

$$k \equiv q_i (mod\ p_i^{\ell_i}) \quad (1 \le i \le n) \tag{6.3.18}$$

has the unique solution $mod\ \Pi_{i=1}^n p_i^{\ell_i}$. It follows that there are as many as

$$\left[\frac{m}{\Pi_{i=1}^n p_i^{\ell_i}}\right] + \theta,$$

$\theta = 0$ or 1, integers in $\{1, 2, \ldots, m\}$ that satisfy congruence system (6.3.18). Write

$$x_i = \sum_{\ell=0}^{\infty} a_{i,\ell} p_i^{-\ell-1} \quad (1 \le i \le n).$$

Then

$$N_m(x_1 \cdots, x_n) = \sum_{\ell_1=1}^{M_1+2} \cdots \sum_{\ell_n=1}^{M_n+2} (\Pi_{i=1}^n b_{i,\ell_i-1}) \left(\left[\frac{m}{\Pi_{i=1}^n p_i^{\ell_i}}\right] + \theta \right),$$

where

$$M_i = \left[\frac{\log m}{\log p_i}\right],$$

$b_{i,\ell} = a_{i,\ell}$ $(0 \le \ell \le M_i)$, and $b_{i,M_i+1} = 1$.

In addition,

$$
\begin{aligned}
&mx_1 \cdots x_n \\
= &\sum_{\ell_1=1}^{\infty} \sum_{\ell_n=1}^{\infty} (\Pi_{i=1}^n a_{i,\ell_i-1}) \left(\left[\frac{m}{\Pi_{i=1}^n p_i^{\ell_i}} \right] + \left\{ \frac{m}{\Pi_{i=1}^n p_i^{\ell_i}} \right\} \right),
\end{aligned}
$$

$\{x\}$ being the fractional part of x; i.e., $\{x\} = x - [x]$. Since for all $q \geq 1$

$$\left[\frac{m}{p_i^{M_i+1} q} \right] = 0,$$

we have

$$
\begin{aligned}
&|N_m(x_1, \ldots, x_n) - mx_1 \cdots x_n| \\
= &\left| \sum_{\ell_1=1}^{M_1+2} \cdots \sum_{\ell_n=1}^{M_n+2} (\Pi_{i=1}^n b_{i,\ell_i-1}) \theta \right. \\
&\left. - \sum_{\ell_1=1}^{\infty} \sum_{\ell_n=1}^{\infty} (\Pi_{i=1}^n a_{i,\ell_i-1}) \left\{ \frac{m}{\Pi_{i=1}^n p_i^{\ell_i}} \right\} \right| \\
\leq &\left| \sum_{\ell_1=1}^{M_1+2} \cdots \sum_{\ell_n=1}^{M_n+2} (\Pi_{i=1}^n b_{i,\ell_i-1}) \theta \right. \\
&\left. - - \sum_{\ell_1=1}^{M_1+1} \sum_{\ell_n=1}^{M_n+1} (\Pi_{i=1}^n a_{i,\ell_i-1}) \left\{ \frac{m}{\Pi_{i=1}^n p_i^{\ell_i}} \right\} \right| \\
&+ \sum_{\nu=1}^{n} \frac{m}{p_\nu^{M_\nu+1}} \sum_{m_1=1}^{\infty} \cdots \sum_{m_n=1}^{\infty} (\Pi_{i=1}^n a_{i,\ell_i-1}) \frac{1}{\Pi_{i=1}^n p_i^{\ell_i}} \\
\leq &\left| \sum_{\ell_1=1}^{M_1+2} \cdots \sum_{\ell_n=1}^{M_n+2} (\Pi_{i=1}^n b_{i,\ell_i-1}) \left(\theta - \left\{ \frac{m}{\Pi_{i=1}^n p_i^{\ell_i}} \right\}' \right) \right| \\
&+ \sum_{\nu=1}^{n} \frac{m}{p_\nu^{M_\nu+1}} \Pi_{i=1}^n \frac{p_i - 1}{p_i \left(1 - \frac{1}{p_i}\right)},
\end{aligned}
$$

where

$$\left\{ \frac{m}{\Pi_{i=1}^n p_i^{\ell_i}} \right\}' = 0$$

when $\ell_i = M_i + 2$. Therefore,

$$
\begin{aligned}
&|N_m(x_1, \ldots, x_n) - mx_1 \cdots x_n| \\
\leq &\Pi_{i=1}^n \left(\sum_{\ell_i=1}^{M_i+2} b_{i,\ell_i-1} \right) + n \leq \Pi_{i=1}^n \left(2 + \sum_{\ell_i=1}^{M_i+1} a_{i,\ell_i-1} \right) \\
\leq &\Pi_{i=1}^n \left((M_i + 1)(p_i - 1) + 2 \right) \leq \Pi_{i=1}^n 2M_i p_i \\
\leq &2^n \left(\Pi_{i=1}^n \frac{p_i}{\log p_i} \right) \log^n m.
\end{aligned}
$$

The theorem is proved. □

From Lemmas 6.3.1 and 6.3.2 and Theorem 6.3.3, we immediately obtain the following result.

Theorem 6.3.4 *Let $N > p_n$ and $f \in H_n^1(C)$. Then the numerical quadrature formula*

$$\int_{V_n} f(X)dX = \frac{1}{N}\sum_{k=1}^{N} f(\phi_{p_1}(k), \dots, \phi_{p_n}(k)) + \rho_N$$

has for its error term the estimate

$$|\rho_N| \leq 4^n C \Pi_{i=1}^n \left(\frac{p_i}{\log p_i}\right) \frac{\log^n N}{N}.$$

Here the error bound cannot be improved for the class as a whole.

As a corollary, we have the following result about the numerical quadrature formula derived at the beginning of this section.

Corollary 6.3.5 *Let $F_i(X)$ be a function defined as (6.3.2). Denote*

$$\begin{aligned} & F(i, M_k; n, N) \\ = \ & \left[F_i\left(\phi_{p_1}(k), \dots, \phi_{p_{i-1}}(k), x_i, \phi_{p_i}(k), \dots, \phi_{p_{n-1}}(k)\right)\right]_{x_i=0}^{x_i=1}. \end{aligned}$$

We have the BTQF

$$\begin{aligned} \int_{V_n} u dV &= \sum_{i=1}^{n} \int_{V_{n-1}} [F_i(X)]_{x_i=0}^{x_i=1} dV_{n-1} \\ &= \frac{1}{N}\sum_{i=1}^{n}\sum_{k=1}^{N} F(i, M_k; n, N) + \rho_N, \end{aligned} \tag{6.3.19}$$

where $N > p_{n-1}$ and the remainder satisfies

$$|\rho_N| \leq n4^{n-1} C \Pi_{i=1}^{n-1} \left(\frac{p_i}{\log p_i}\right) \frac{\log^{n-1} N}{N}.$$

C is a constant and this estimate cannot be improved for the class as a whole.

6.4 Applications of DREs in the boundary element method

In this section, we will give an application of the DREs, (6.2.6) and (6.2.8), developed in Section 6.2, in the *boundary element method*, a popular method for solving partial differential equations numerically.

Let us consider the following boundary value problem

$$\begin{cases} Lu(X) = g(X) & X \in \Omega, \\ u = \overline{u} & X \in \partial\Omega_1, \\ g = \frac{\partial u}{\partial n} = \overline{g} & X \in \partial\Omega_2, \end{cases} \tag{6.4.1}$$

where $\partial\Omega_1 \cup \partial\Omega_2 = \partial\Omega$. The solution process, as shown below, is analogous to the *boundary element method (BEM)*.

The BEM is used for evaluating the numerical solution of a boundary value problem of a partial differential equation. It essentially consists of three steps: (i) establishing the boundary integral equation for the given problem; (ii) solving the boundary integral equation using finite element method techniques; and (iii) finding the approximate solution of the problem in the interior of the domain by using the boundary solution obtained in step (ii) and by applying Green's formula (cf. [1], [5], [34], [35], [72], [73], [74]). So, the BEM only needs to discretize the boundary of the domain, and this requires very simple data input and storage techniques. If the BEM is used to solve an *exterior problem* (e.g., problem (6.4.1) with the domain Ω such that $R \setminus \Omega$ is bounded), it is not necessary to deal with the boundary at infinity, since the corresponding fundamental solution chosen in the BEM satisfies the radiation condition.

Thus, exterior problems with unbounded domains are handled as easily as *interior problems*. That is, the BEM is much more suitable for problems on unbounded domains. We will show that a method derived from the DREs in Section 6.2 maintains all of these advantages. Moreover, a spline approximation will be applied in the discretization of the DREs, so that hypersingular integrals in the DREs can be reduced to weakly singular integrals. In contrast to classic BEM, the point collocation procedures are always used in step (ii). In these procedures, the unknown function and its normal derivative are approximated by individual piecewise continuous linear functions or step functions. Thus, hypersingular integrals must be treated using some smoothness approximation, which produce large errors.

We now derive a method for the boundary value problem (6.4.1). Let $u \in C(\bar{\Omega})$ be the solution of $L(u) = g$ and v be a *fundamental solution* of $Mv = 0$; i.e., a solution of $Mv = \delta(X - X_0)$. Here M is the adjoint differential operator of L, $\delta(X)$ is the Delta function, and X_0, called the source point, is an arbitrarily fixed point in Ω. In general, the fundamental solution exists but is not unique. We now evaluate $\int_\Omega uMvdX$ for the fundamental solution v. If $X_0 \in \Omega\backslash\partial\Omega$, then

$$\int_\Omega u(X)Mv(X)dX = \int_\Omega u(X)\delta(X - X_0)dX = u(X_0).$$

DRE (6.2.8) yields the following equation of $u = u(X)$.

$$\begin{aligned} u(X_0) &= \int_{\Omega} vg\, dX + \int_{\partial\Omega} \left[u \sum_{i=1}^{n} p_i \frac{\partial x_i}{\partial \nu} \right] dS \\ &- \int_{\partial\Omega} \left[v \sum_{i=1}^{n} \left(\sum_{j=1}^{n} a_{ij}(X) \frac{\partial u}{\partial x_j} \right) \frac{\partial x_i}{\partial \nu} \right] dS. \end{aligned} \tag{6.4.2}$$

We now give an application of formula (6.4.2) to harmonic equations. Let us first consider the plane harmonic equation, which has the elliptic operator L shown by (6.2.2) with $n = 2$, $a_{ij} = \delta_{ij}$, and all b_i and c being zero. In this case, L is also denoted by Δ, and, obviously, the adjoint operator $M = L = \Delta$. In the polar coordinate system, the fundamental solution, $v = v(r)$, of Δ satisfies $\Delta v = \delta(X)$:

$$\left(\frac{\partial^2}{\partial r^2} + \frac{1}{r} \frac{\partial}{\partial r} \right) v(r) = \delta(X).$$

Thus,

$$\frac{1}{r} \frac{d}{dr} \left(r \frac{dv}{dr} \right) = \delta(X).$$

Integrating on both sides of the above equation over the circular domain $|X| \leq r$ $(r > 0)$ yields

$$2\pi r \frac{dG}{dr} = 1.$$

It follows that

$$v(r) = \frac{1}{2\pi} \log r.$$

In DRE (6.2.20), taking u as the solution of $\Delta u = 0$ and $v = \log|X - X_0|/2\pi$ while noting that

$$\frac{\partial f}{\partial \nu} = \sum_{i=1}^{n} \frac{\partial f}{\partial x_i} \frac{\partial x_i}{\partial \nu},$$

we obtain

$$\begin{aligned} &u(X_0) \\ = \;& \tfrac{1}{2\pi} \textstyle\int_{\partial\Omega} \left[u(X) \tfrac{\partial}{\partial \nu} \log|X - X_0| - \log|X - X_0| \tfrac{\partial u(X)}{\partial \nu} \right] dS. \end{aligned}$$

Similarly, for harmonic equations with dimension $n \geq 3$, the corresponding fundamental solution is

$$v(X) = -\frac{1}{(n-2)r^{n-2}A_{n-1}},$$

where $r = |X|$ and A_{n-1} is the area of the unit sphere in $\mathbb{R}^n$. For example, when $n = 3$, $v(X) = -(4\pi r)^{-1}$. In this case, taking u as the solution of $\Delta u = 0$ and $v = v(|X - X_0|) = -(4\pi|X - X_0|)^{-1}$ in DRE (6.2.20),

$$\begin{aligned} & u(X_0) \\ = & \frac{1}{4\pi}\int_{\partial\Omega}\left[|X - X_0|^{-1}\frac{\partial u(X)}{\partial \nu} - u(X)\frac{\partial |X - X_0|^{-1}}{\partial \nu}\right] dS. \quad (6.4.3)\end{aligned}$$

If the values of $u(X)$ and of its derivatives on the boundary $\partial\Omega$ are obtained, then expression (6.4.3) can be applied to evaluate all values of $u(X_0)$ in the interior of Ω. However, in boundary value problem (6.4.1) we are given only some of the values of $u(X)$ and its derivatives on $\partial\Omega$. Hence, we need to calculate those missing values on the boundary. In the following, we will use the last example (i.e., harmonic equation with dimension $n = 3$) to illustrate the process of finding the unknown values.

We first consider boundary value problem (6.4.1) with the operator $L = \Delta$ and a bounded domain Ω; i.e., an interior problem of the Laplace equation. While evaluating the missing values on the boundary, we cannot apply DRE formula (6.2.8) directly on Ω because of the singularity of the fundamental solution $v = v(|X - X_0|) = -(4\pi|X - X_0|)^{-1}$ at the point X_0 on $\partial\Omega$. Therefore, we consider a small ball $B_\epsilon = \{X : |X - X_0| < \epsilon\}$ and apply DRE (6.2.8) on the domain $\Omega_\epsilon = \Omega \setminus \overline{B_\epsilon}$ for the operators $L = M = \Delta$. It follows that

$$\int_{\Omega_\epsilon}(u\Delta v - v\Delta u)\, dX = \int_{\partial\Omega_\epsilon}\left[u\frac{\partial v}{\partial \nu} - v\frac{\partial u}{\partial \nu}\right] dS. \quad (6.4.4)$$

Substituting the solution of $\Delta u = 0$, $u \in C(\bar{\Omega})$, and $v = -(4\pi|X - X_0|)^{-1}$ into equation (6.4.4), we have $\Delta v = 0$ for all $X \in \Omega_\epsilon$ and obtain

$$\begin{aligned} 0 &= \frac{1}{4\pi}\int_{\partial\Omega_\epsilon}\left[|X - X_0|^{-1}\frac{\partial u(X)}{\partial \nu} - u(X)\frac{\partial |X - X_0|^{-1}}{\partial \nu}\right] dS \\ &= I_\epsilon^{(1)} + I_\epsilon^{(2)}, \quad (6.4.5)\end{aligned}$$

where

$$I_\epsilon^{(1)} = \frac{1}{4\pi}\int_{\partial\Omega_\epsilon\setminus K_\epsilon}\left[|X - X_0|^{-1}\frac{\partial u(X)}{\partial \nu} - u(X)\frac{\partial |X - X_0|^{-1}}{\partial \nu}\right] dS, \quad (6.4.6)$$

$K_\epsilon = \Omega \cap \partial B_\epsilon$, and

$$I_\epsilon^{(2)} = \frac{1}{4\pi} \int_{K_\epsilon} \left[|X - X_0|^{-1} \frac{\partial u(X)}{\partial \nu} - u(X) \frac{\partial |X - X_0|^{-1}}{\partial \nu} \right] dS. \tag{6.4.7}$$

Denote by $d\theta$ the solid angle of K_ϵ with respect to X_0. Then equation (6.4.7) becomes

$$I_\epsilon^{(2)} = \frac{1}{4\pi} \int_{K_\epsilon} \left(\epsilon^{-1} \frac{\partial u(X)}{\partial \nu} - \epsilon^{-2} u(X) \right) \epsilon^2 d\theta.$$

If the plane tangent to $\partial\Omega$ at X_0 exists, then the solid angle of K_ϵ tends to 2π as $\epsilon \to 0$. Hence, from the mean value theorem for an integral, we obtain

$$\lim_{\epsilon \to 0} I^{(2)} = -\frac{1}{2} u(X_0).$$

Taking the limit as $\epsilon \to 0$ in equation (6.4.5) yields the integral equation

$$\begin{aligned} & u(X_0) \\ = \; & \frac{1}{2\pi} \int_{\partial\Omega} \left(|X - X_0|^{-1} \frac{\partial u(X)}{\partial \nu} - u(X) \frac{\partial |X - X_0|^{-1}}{\partial \nu} \right) dS \end{aligned} \tag{6.4.8}$$

for all $X_0 \in \partial\Omega$.

Obviously, if there does not exist a plane tangent to $\partial\Omega$ at X_0, then we replace the 2π on the right-hand side of (6.4.8) by the corresponding solid angle of Ω with respect to point X_0 and obtain the corresponding integral equation. In addition, if $X_0 \in \mathbb{R}^3 \setminus \Omega$, then $\Delta v = \Delta |X - X_0|^{-1} = 0$ for all $X \in \bar{\Omega}$. Therefore,

$$\int_{\partial\Omega} \left(|X - X_0|^{-1} \frac{\partial u(X)}{\partial \nu} - u(X) \frac{\partial |X - X_0|^{-1}}{\partial \nu} \right) dS = 0 \tag{6.4.9}$$

for all $X_0 \in \mathbb{R}^3 \setminus \Omega$. When $X_0 \in \Omega$ and $X_0 \in \partial\Omega$, we have expression (6.4.3) of $u(X_0)$ and the integral equation (6.4.8), respectively.

Next, we consider boundary value problem (6.4.1) with the operator $L = \Delta$ and a domain Ω whose complement $\Omega' = \mathbb{R}^3 \setminus \bar{\Omega}$ is bounded. This boundary value problem is called the exterior problem of the Laplace equation. In order to obtain the unique solution to this problem, we also need for u to satisfy the radiation conditions

$$u(X) = O\left(\frac{1}{|X|}\right), \quad \|grad\ u(X)\| = O\left(\frac{1}{|X|^2}\right) \tag{6.4.10}$$

as $|X| \to \infty$. We now derive expressions of $u(X_0)$ for $X_0 \in \Omega$ and $X_0 \in \partial\Omega$. Let $B_r = \{X : X \in \mathbb{R}^3, |X| \leq r\}$ be a large enough ball that contains all of Ω'. Denote $\Omega_r = \Omega \cap B_r$. Applying formula (6.4.3) to the domain Ω_r gives

$$\begin{aligned} & u(X_0) \\ = & \frac{1}{4\pi}\int_{\partial\Omega}\left[|X-X_0|^{-1}\frac{\partial u(X)}{\partial \nu} - u(X)\frac{\partial |X-X_0|^{-1}}{\partial \nu}\right] dS \\ + & \frac{1}{4\pi}\int_{\partial B_r}\left[|X-X_0|^{-1}\frac{\partial u(X)}{\partial \nu}\right. \\ - & \left. u(X)\frac{\partial |X-X_0|^{-1}}{\partial \nu}\right] dS. \end{aligned} \tag{6.4.11}$$

Noting the uniqueness conditions (6.4.10), we find that the absolute value of the last integral of equation (6.4.11) is bounded by

$$\begin{aligned} & \frac{1}{4\pi}\int_{\partial B_r}\left[|u(X)|\left|\frac{\partial |X-X_0|^{-1}}{\partial \nu}\right| + |X-X_0|^{-1}\left|\frac{\partial u(X)}{\partial \nu}\right|\right] dS \\ \leq & \frac{C}{4\pi r} \to 0 \end{aligned}$$

as $r \to \infty$. Therefore, by taking the limit $r \to \infty$ on equation (6.4.11), we have

$$\begin{aligned} & u(X_0) \\ = & \frac{1}{4\pi}\int_{\partial\Omega}\left[|X-X_0|^{-1}\frac{\partial u(X)}{\partial \nu}\right. \\ - & \left. u(X)\frac{\partial |X-X_0|^{-1}}{\partial \nu}\right] dS \end{aligned} \tag{6.4.12}$$

for all $X_0 \in \Omega$. Similar to the interior problem, we can obtain

$$\begin{aligned} & u(X_0) \\ = & \frac{1}{2\pi}\int_{\partial\Omega}\left[|X-X_0|^{-1}\frac{\partial u(X)}{\partial \nu}\right. \\ - & \left. u(X)\frac{\partial |X-X_0|^{-1}}{\partial \nu}\right] dS \end{aligned} \tag{6.4.13}$$

for all $X_0 \in \partial\Omega$, and

$$\int_{\partial\Omega}\left(|X-X_0|^{-1}\frac{\partial u(X)}{\partial \nu} - u(X)\frac{\partial |X-X_0|^{-1}}{\partial \nu}\right) dS = 0 \tag{6.4.14}$$

for all $X_0 \in \Omega'$.

We now give the BEM procedure for solving boundary value problem (6.4.1) using DRE (6.2.8). Similar to the examples shown above, we use the limit process on DRE (6.2.8) and obtain

$$\begin{aligned} \alpha u(X_0) &= \int_\Omega vg\, dX + \int_{\partial\Omega} \left[u \sum_{i=1}^n p_i \frac{\partial x_i}{\partial \nu} \right] dS \\ &- \int_{\partial\Omega} \left[v \sum_{i=1}^n \left(\sum_{j=1}^n a_{ij}(X) \frac{\partial u}{\partial x_j} \right) \frac{\partial x_i}{\partial \nu} \right] dS. \end{aligned} \tag{6.4.15}$$

In this expression, $\alpha = 1$ if X_0 is in the interior of Ω, and it is a positive real number less than 1 if $X_0 \in \partial\Omega$ (in particular, $\alpha = \frac{1}{2}$ if X_0 is on the smooth boundary). Thus, we obtain an equation about $u(X_0)$ and the boundary type weighted integrals of u and $\frac{\partial u}{\partial n}$. Here $\frac{\partial u}{\partial n} = \sum_{i=1}^n \frac{\partial u}{\partial x_i} \frac{\partial x_i}{\partial \nu}$ is the outward normal derivative of u on $\partial\Omega$. Applying a quadrature formula (e.g., the formula given in Section 3) to these boundary integrals, we obtain an algebraic equation, called the basic algebraic equation, about $u(X_0)$ and the values of u and $\frac{\partial u}{\partial n}$ at the nodes on $\partial\Omega$. Replacing the source point X_0 with each node of the boundary quadrature formula in the basic algebraic equation, we generate a system of linear equations about the values of u and $\frac{\partial u}{\partial n}$ at the nodes on $\partial\Omega$. Substituting the given boundary conditions, which are the given values of u and $\frac{\partial u}{\partial n}$ at some nodes on $\partial\Omega$, we can solve for the unknown values of u and $\frac{\partial u}{\partial n}$ at the other nodes on $\partial\Omega$. After finding all the boundary values, the value of u at any interior point X_0 in Ω can be evaluated using a basic algebraic equation with the values of u and $\frac{\partial u}{\partial n}$ at the nodes on $\partial\Omega$.

As an example, we consider the following boundary value problem of *Helmholtz's equation,*

$$\begin{cases} (\nabla^2 + k)u = 0 & \text{in } \Omega, \\ u = \overline{u} & \text{on } \partial\Omega_1, \\ q = \frac{\partial u}{\partial n} = \overline{q} & \text{on } \partial\Omega_2, \end{cases} \tag{6.4.16}$$

where $\nabla^2 u = \frac{\partial^2 u}{\partial x^2} + \frac{\partial^2 u}{\partial y^2}$, $\partial\Omega_1 \cup \partial\Omega_2 = \partial\Omega$, and $\partial\Omega$ is piecewise smooth. Obviously, $L = \nabla^2 + k$ is a self-adjoint operator. Therefore, $M = L = \nabla^2 + k$. Assume that a partition on the boundary $\partial\Omega$ is given by $\{X_i = (x_i, y_i) = (x(t_i), y(t_i))\}_{i=1}^n$, where $(x(t), y(t))$ is a parametric form of the boundary. We will find all unknown values of $u_i = u(x_i, y_i)$ and $\frac{\partial u}{\partial n}(x_i, y_i)$. For any fixed point X_i (called the source point), the corresponding fundamental solution of $Mv = 0$ (i.e., the solution of $Mv = \delta(X - X_i)$) is $v = v(X, X_i) = -\frac{i}{4} H_0^{(2)}(kr)$; here $H_0^{(2)}$ is the zeroth order *Hankel function* of the second kind, and $r = |X - X_i|$ denotes the distance from a point $X = (x, y)$ to the source point $X_i = (x_i, y_i) \in \partial\Omega$. From (6.4.15), observing $g = 0$ and $a_{ij} = \delta_{ij}$, $1 \le i, j \le 2$, we have

$$\left(1 - \frac{\beta}{2\pi}\right) u(x_i, y_i) = \int_{\partial\Omega} u \frac{\partial v}{\partial n}\, dS - \int_{\partial\Omega} v \frac{\partial u}{\partial n}\, dS, \tag{6.4.17}$$

where β is defined as follows. Two lines tangent to $\partial\Omega$ originate from the source point (x_i, y_i). β is just the measure of the angle between the two lines that traverse Ω. In particular, if $\partial\Omega$ is smooth at (x_i, y_i), then $\beta = \pi$. In (6.4.17), since $v = v(x, y; x_i, y_i) = -\frac{j}{4}H_0^{(2)}(kr)$, $r = |(x, y) - (x_i, y_i)|$, we have

$$\begin{aligned}\frac{\partial v}{\partial n} &= \frac{\partial v}{\partial n}(x, y; x_i, y_i) \\ &= \frac{j}{4}kH_1^{(2)}(kr)\cos(\vec{r}, \vec{n}),\end{aligned} \tag{6.4.18}$$

where $\vec{r} = (x - x_i, y - y_i)$, $\vec{n}$ is the outward normal vector at $(x, y) \in \partial\Omega$, and $H_1^{(2)}$ is the first order Hankel function of the second kind. Assume $\partial\Omega$ is defined by the parametric function $c(t) = (x(t), y(t))$, $a \le t \le b$. We denote $(x_j, y_j) = (x(t_j), y(t_j))$, $u_j = u(x_j, y_j) = u(x(t_j), y(t_j))$, $q_j = \frac{\partial u}{\partial n}(x_j, y_j) = \frac{\partial u}{\partial n}(x(t_j), y(t_j))$, and $t_j = (b - a)\frac{j}{n}$, $j = 1, 2, \ldots, n$. To discretize the integrals in equation (6.4.17), we use interpolation to expand u and g approximately as

$$u \doteq \sum_{j=1}^{n} u_j\phi_j^{(1)}(t) \text{ and } q \doteq \sum_{j=1}^{n} q_j\phi_j^{(2)}(t), \tag{6.4.19}$$

where $\phi_j^{(k)}(t)$, $k = 1, 2$, are the Lagrange interpolation basis functions. We can choose as $\phi_j^{(k)}(t)$, $k = 1, 2$, either the Lagrange basis function $l_j(t) = \prod_{i=1, i\neq j}^{n} \frac{t - t_i}{t_j - t_i}$ (see Section 2.3) or some other basis function. For instance, we can assume that $\phi_j^{(k)}(t) = \phi^{(k)}(t - j/n)$, $j = 1, \ldots, n$, and $\{\phi^{(k)}(2nt - 2j)\}_{j\in Z}$, $k = 1, 2$, are the basis functions of the optimal spline Lagrange interpolation for the data at the even integers [29]. Here an *optimal spline interpolation* is an interpolation with a spline basis function that possesses the highest possible approximation order and the smallest possible compact support. For example, we can choose

$$\phi^{(2)}(\frac{t}{2h}) = N_2(t + 1) \equiv (1 - |t|)\chi_{[-1,1]}(t)$$

and

$$\phi^{(1)}(\frac{t}{h}) = \frac{3}{2}\phi(t) + \frac{1}{2}[\phi(t - 1) + \phi(t + 1)] - \frac{1}{4}[\phi(t - 2) + \phi(t + 2)], \tag{6.4.20}$$

where $N_2(t)$ is the *B-spline* of order 2, $h = (b - a)/(2n)$, $\phi(t) = [N_2(t + 1) + N_2(t + 2)]/2$, $supp\ \phi^{(1)} = [-4, 4]$, and $supp\ \phi^{(2)} = [-2, 2]$. It is easy to check that $\phi^{(1)}(i/n) = \delta_{i0}$ and that the corresponding interpolation on the interval $[a, b]$

with the basis $\{\phi_j^{(1)}(t)\}$ has the optimal approximation order of $O((n)^{-2})$. $\phi^{(2)}$ also satisfies $\phi^{(2)}(i/n) = \delta_{i0}$. If we need more approximation accuracy, we can choose higher smooth spline functions for $\phi^{(1)}$, such as

$$\begin{aligned}\phi^{(1)}(\frac{t}{h}) &= -\frac{1}{8}N_3(t-1) + \frac{1}{8}N_3(t) + N_3(t+1) \\ &+ N_3(t+2) + \frac{1}{8}N_3(t+3) - \frac{1}{8}N_3(t+4),\end{aligned} \tag{6.4.21}$$

where $h = (b-a)/(2n)$ and $supp\ \phi^{(1)} = [-4, 4]$. Clearly, $\phi^{(1)}(i/n) = \delta_{i0}$, $\phi^{(2)}(i/n) = \delta_{i0}$, and the interpolation corresponding to $\phi^{(1)}(i/n)$ has the optimal approximation order of $O((n)^{-3})$.

Now we consider the two-dimensional scattering problem of electromagnetic wave incident on an infinitely long circular conducting body with radius a (cf. [78], [50], and [7]). We express any point $(x, y) \in \partial\Omega$ parametrically as follows. The parameter θ is the angle formed in clockwise orientation from the positive x-axis to the ray from the origin to the point (x, y), and

$$x = a\cos(\theta), \quad y = a\,sin(\theta).$$

Thus, the nodes $(x_i, y_i) \in \partial\Omega$ are $x_i = a\ cos(\theta_i)$, and $y_i = a\ sin(\theta_i)$, where $\theta_i = -2\pi i/n$, $i = 1, \ldots, n$. Also,

$$\begin{aligned} r &= \sqrt{(x-x_i)^2 + (y-y_i)^2} \\ &= 2a\left|\sin\frac{\theta-\theta_i}{2}\right|, \\ \cos(\mathbf{r}, \mathbf{n}) &= -\left|\sin\frac{\theta-\theta_i}{2}\right|. \end{aligned}$$

We apply (6.4.19) to give the approximate expressions of u and q as

$$u = \sum_{j=1}^{n} u_j\phi_j^{(1)}(\theta), \quad q = \sum_{j=1}^{n} q_j\phi_j^{(2)}(\theta), \tag{6.4.22}$$

where $\phi_j^{(k)}(\theta) = \phi^{(k)}(\theta - \theta_j)$ $(k = 1, 2)$, $\phi^{(1)}(\theta/h)$ is as was shown in (6.4.20), $\phi^{(2)}(\theta/(2h)) = N_2(t+1)$, and $h = \frac{\pi}{n}$.

Substituting (6.4.22) into (6.4.17) while noting that $\beta = \pi$, we obtain

$$\frac{1}{2}u_i - \sum_{j=1}^{n} u_j \int_0^{2\pi} \phi_j^1(\theta)\frac{\partial v}{\partial n}(\theta, \frac{i}{n})\,ds(\theta) = -\sum_{j=1}^{n} q_j \int_0^{2\pi} \phi_j^2(\theta)v(\theta, \frac{i}{n})\,ds(\theta), \tag{6.4.23}$$

where

$$v(\theta, \frac{i}{n}) = v(x(\theta), y(\theta); x(2\pi\frac{i}{n}), y(2\pi\frac{i}{n}))$$

and

$$\frac{\partial v}{\partial n}(\theta, \frac{i}{n}) = \frac{\partial v}{\partial n}(x(\theta), y(\theta); x(2\pi\frac{i}{n}), y(2\pi\frac{i}{n})).$$

Discretizing the integrals in (6.4.17) yields system (6.4.23). To solve the system, we substitute $v = -\frac{j}{4}H_0^{(2)}(kr)$ and $\frac{\partial v}{\partial n} = \frac{j}{4}kH_1^{(2)}(kr)\cos(\vec{r}, \vec{n})$ into it and compute

$$h_{ij} = -\int_0^{2\pi} \phi_j^{(1)}(\theta)\frac{\partial v}{\partial n}(\theta, \frac{i}{n})\, ds(\theta) + \frac{\delta_{ij}}{2} \qquad (6.4.24)$$

and

$$g_{ij} = -\int_0^{2\pi} \phi_j^{(2)}(\theta)v(\theta, \frac{i}{n})\, ds(\theta). \qquad (6.4.25)$$

Thus, (6.4.23) can be written as

$$\mathbf{Hu} = \mathbf{Gq}, \qquad (6.4.26)$$

where $\mathbf{H} = [h_{ij}]_{1\le i,j\le n}$, $\mathbf{G} = [g_{ij}]_{1\le i,j\le n}$, $\mathbf{u} = (u_1, u_2, \dots, u_n)^T$, and $\mathbf{q} = (q_1, q_2, \dots, q_n)^T$. Substituting the given boundary conditions of the problem (6.4.1) into (6.4.26), we can solve the linear system (6.4.23) for all unknown u_j and q_j.

To evaluate (6.4.24) and (6.4.25), we need the following asymptotes of $H_k^{(2)}$ $(k = 1, 2)$.

$$H_0^{(2)}(k\epsilon) = 1 - \frac{2}{\pi}j\,(\ln(k\epsilon) + \gamma - \ln 2) + O\left(\epsilon^2 \ln \epsilon\right) \qquad (6.4.27)$$

and

$$\begin{aligned} H_1^{(2)}(k\epsilon) \quad &= \quad \frac{k\epsilon}{2}\left[1 - \frac{2}{\pi}j\left(\ln(k\epsilon) + \gamma - \frac{3}{4} - \ln 2\right)\right] \\ &\quad + \frac{1}{2k\epsilon\pi}j + O\left(\epsilon^3 \ell n\epsilon\right) \end{aligned} \qquad (6.4.28)$$

as $\epsilon \to 0$. Here γ is *Euler's number.*

In the calculation of (6.4.25), we must treat the singular integrals

$$\int_0^1 H_0^{(2)}(cu)du \qquad (6.4.29)$$

and

$$\int_0^1 H_0^{(2)}(cu)u du. \tag{6.4.30}$$

By using asymptote (6.4.27), integral (6.4.30) can be evaluated as

$$\lim_{\epsilon \to 0} \frac{1}{c} u H_1^{(2)}(cu)\Big|_\epsilon^1 = \frac{1}{c} H_1^{(2)}(c) - \frac{j}{4c^2\pi}.$$

Similarly, integral (6.4.29) is equal to

$$\begin{aligned} & \frac{1}{c}\int_0^c H_0^{(2)}(t)dt \\ = & \frac{1}{c}\left[tH_0^{(2)}(t) + \frac{\pi}{2}t\left(H_1^{(2)}(t)S_0(t) - H_0^{(2)}(t)S_1(t)\right)\right]_0^c \\ = & H_0^{(2)}(c) + \frac{\pi}{2}(H_1^{(2)}(c)S_0(c) - H_0^{(2)}(c)S_1(c)), \end{aligned}$$

where S_0 and S_1 are *Struve functions* defined by

$$\begin{aligned} S_0(t) &= 2\pi\left[t - \frac{t^3}{1^2 3^2} + \frac{t^5}{1^2 3^2 5^2} - \frac{t^7}{1^2 3^2 5^2 7^2} + \cdots\right] \\ S_1(t) &= \frac{2}{\pi}\left[\frac{t^2}{1^2 3} - \frac{t^4}{1^2 3^2 5} + \frac{t^6}{1^2 3^2 5^2 7} - \frac{t^7}{1^2 3^2 5^2 7^2 9} + \cdots\right]. \end{aligned}$$

To evaluate integral (6.4.24), we make use of the C^1 property of $\phi_j^{(1)}(\theta)$ to change the following *strong singular integral* into a *weak singular integral.* The hyper-singular integral term in (6.49) is

$$\begin{aligned} & \int_0^{2\pi} \phi_j^{(1)}(\theta)\frac{\partial v}{\partial n}(\theta, \frac{i}{n})\, ds(\theta) \\ = & \frac{ak}{4}\sqrt{-1}\int_0^{2\pi} \phi_j^{(1)}(\theta)\left|\sin\frac{\theta - \theta_i}{2}\right| H_1^{(2)}\left(2ak\left|\sin\frac{\theta-\theta_i}{2}\right|\right) d\theta, \end{aligned}$$

where $j = i-3, i-2, i-1, i, i+1, i+2, i+3$. Since $\phi_j^{(1)}(\theta)$ are in C^1, the above integrals can be changed to some explicit functions plus a linear combination of weak singular integrals of the form (6.4.24) and (6.4.25). Thus, they can be found easily.

In the collocation process of the BEM for the problem of plane wave scattering by an infinite conducting cylinder, employing the C^1 quadratic finite element $\psi^{(1)}$ (see (6.4.20)) as a basis compares favorably with an application of linear elements. For example, let the number of nodes, n, be 60, and so the segment length, $\Delta\theta_j = \theta_{j+1} - \theta_j$, is approximately $0.1a$, where a is the radius of the intersection circle of the cylinder. In this case, [7] has shown that the solution calculated using C^1 quadratic finite elements has an average amplitude error smaller than that from using linear elements.

Using the C^1 quadratic finite elements as a basis for the collocation process in BEM has several other advantages: (1) it is not necessary to map an arbitrary segment of the boundary into a standard segment where the shape function is defined over a given coordinate system; (2) the smoothness of the solution is guaranteed because the basis function is smooth everywhere; (3) strong singularities can be reduced to weak singularities so that numerical integration can be performed easily; (4) the order of approximation is known a priori (cf. [73]) so that the mesh size can be predetermined for any specified required accuracy.

Finally, after obtaining all values of u and $\partial u/\partial n$ on the boundary $\partial\Omega$, we use the values to evaluate the value of u at any point (x_0, y_0) in the interior of Ω from (6.4.17) for $\beta = 0$ or, equivalently, from

$$\begin{aligned} u(x_0, y_0) &= \sum_{j=1}^{n} u_j \int_0^{2\pi} \phi_j^1(t) \frac{\partial v}{\partial n}(t, t^0)\, ds(t) \\ &\quad - \sum_{j=1}^{n} q_j \int_0^{2\pi} \phi_j^2(t) v(t, t^0)\, ds(t). \end{aligned}$$

Thus we obtain the numerical solution of u on the boundary $\partial\Omega$ and at any interior point of Ω.

At the conclusion of this section, we note that a wavelet technique can be utilized to evaluate the boundary integrals in (6.4.23). First, expand v and $\frac{\partial v}{\partial n}$ in terms of $\psi_{mk}^*(t)$, the periodized version of $\psi_{mk}(t) = 2^{m/2}\psi(2^m - k)$. $\psi(t)$ is the wavelet associated with the scaling function $\phi(t) = \phi^{(1)}(nt)$ shown in (6.4.21) (see [29] and [31]). Hence, $\psi(t) = \sum_k \tilde{q}_k \phi(2t - k)$, where the coefficients $\{\tilde{q}_k\}$ are determined by their two-scale symbol $\tilde{Q}(z) = \frac{1}{2}\sum_k \tilde{q}_k z^k$, $z = e^{-iw/2}$. From two papers by the author, [29] and [31], we have $\tilde{Q}(z) = \frac{c(-z)}{c(z^2)} Q(z)$. Here $c(z) = -\frac{1}{8}z^{-1} + \frac{1}{8}z^0 + z^1 + z^2 + \frac{1}{8}z^3 - \frac{1}{8}z^4$, $Q(z) = \frac{1}{2}\sum_k q_k z^k$, $z = e^{-iw/2}$, and

$$q_k = \begin{cases} \frac{(-1)^k}{4} \sum_{l=0}^{3} \binom{3}{l} N_6(k+1-l), & 0 \le k \le 7, \\ 0, & \text{otherwise} \end{cases}$$

is the two-scale symbol of the B-wavelet of order 3 (see [9]). More details can be found in [29] and [31].

Bibliography

[1] P.K. Banerjee and R. Butterfield, *Boundary Element Methods in Engineering Science*, McGraw-Hill Book Company (UK) Limited, New York, 1981.

[2] B. Bradie, R. Coifman, and A. Grossmann, *Fast numerical computations of oscillatory integrals related to acoustic scattering, I,* Appl. Compu. Harmonic Anal., **1**(1993), 94–99.

[3] C.A. Brebbia, ed., *Boundary Element Techniques in Computer-Aided Engineering*, Martinus Nijhoff Publishers, Dordrecht, The Netherlands, 1984.

[4] C.A. Brebbia, ed., *Boundary Element Methods*, Proc. of the 8th Intern. Conf. on BEM, Tokyo 1986, Springer-Verlag, Berlin, 1986.

[5] C.A. Brebbia, J.C.F. Telles, and L.C. Wrobel, *Boundary Element Techniques*, Springer-Verlag, Berlin, 1984.

[6] B.L. Burrows, *A new approach to numerical integration,* J. Inst. Math. Applics., **26**(1980), 151–173.

[7] A.K. Chan, C.K. Chui, and T.X. He, *Application of generalized vertex splines to boundary element method for electromagnetic scattering,* Sixth Annual Review of Progress in Applied Computational Mathematics, Monterey, CA, March 19–22, 1990, 329–337.

[8] E.W. Cheney and T.H. Southard, *A survey of methods for rational approximation, etc.,* SIAM Review, **5**(1963), 219–231.

[9] C.K. Chui, *An Introduction to Wavelets*, Academic Press, New York, 1992.

[10] B. Cipra, *What's Happening in the Mathematical Sciences*, Amer. Math. Soc., Providence, RI, 1996.

[11] I. Daubechies, *Ten Lectures on Wavelets*, SIAM, Philadelphia, 1992.

[12] P.J. Davis, *Interpolation and Approximation*, Blaisdell, New York, 1963.

[13] P.J. Davis, *Triangle formulas in the complex plane,* Math. Comp., **18**(1964), 569–577.

[14] P.J. Davis, *Simple quadratures in the complex plane,* Pacific J. Math., **15**(1965), 813–824.

[15] P.J. Davis, *Double integrals expressed as single integrals or interpolatory functions,* J. Approx. Theory, **5**(1972), 276–307.

[16] P.J. Davis and H. Pollak, *On the analytic continuation of mapping functions,* Trans. Amer. Math. Soc., **87**(1958), 198–225.

[17] P.J. Davis and P. Rabinowitz, *Ignoring the singularity in numerical integration,* J. SIAM Ser. B. Numerical Anal., **2**(1965), 367–383.

[18] P.J. Davis and P. Rabinowitz, *Numerical Integration*, Blaisdell, New York, 1967.

[19] P.J. Davis and P. Rabinowitz, *Methods of Numerical Integration*, Academic Press, New York, 1975.

[20] B. Epstein, *Partial Differential Equations*, McGraw-Hill, New York, 1970.

[21] J.D. Federenko, *A formula for the approximate evaluation of double integrals,* Dopovidi. Akad. Nauk Ukrain. RSR, (1964), 1000–1005.

[22] A. Ghizzetti and A. Ossicini, *Quadrature Formula*, Academic Press, New York, 1970.

[23] W. Gröbner, *Über die Konstruktion von Systemen orthogonaler Polynome in ein-und-zwei dimensionalen Bereiche,* Monatsh. Math., **52**(1948), 48–54.

[24] E. Grosswald, *On the integration scheme of Maréchal*, Proc. Amer. Math. Soc., **2**(1951), 706–709.

[25] P.C. Hammer, A.W. Wymore, and A.H. Stroud, *Numerical evaluation of multiple integrals, I,II,* Math. Tables Aids Comput., **11**(1957), 59–67; **12**(1958), 272–280.

[26] T. Havie, *Remarks on an expansion for integrals of rapidly oscillating functions,* BIT, **13**(1973), 16–29.

[27] T.X. He, *Boundary-type quadrature formulas without derivative terms,* J. Math. Res. Expo., **2**(1981), 93–102.

[28] T.X. He, *On the algebraic method for constructing the boundary-type quadrature formulas,* Comp. Math. (China), (1985), No. 1, 1–5.

[29] T.X. He, *Spline interpolation and its wavelet analysis,* Proceedings of the Eighth International Conference on Approximation Theory, C.K. Chui and L.L. Schumaker (eds.), World Scientific Publishing Co., Inc., 1995, 143–150.

[30] T.X. He, *Construction of boundary quadrature formulas using wavelets,* Wavelet Applications in Signal and Image Processing III, SPIE-The International Society for Optical Engineering, A.F. Laine and M.A. Unser (eds.) 1995, 825–836.

[31] T.X. He, *Short time Fourier transform, integral wavelet transform, and wavelet functions associated with splines,* J. Math. Anal. & Appl., **224**(1998), 182–200.

[32] T.X. He, *Boundary quadrature formulas and their applications*, Handbook of Analytic-Computational Methods in Applied Mathematics, G. Anastassiou (ed.), Chapman & Hall/CRC, New York, 2000, 773–800.

[33] E. Hernández and G. Weiss, *A First Course on Wavelets*, CRC Press, New York, 1996.

[34] G.C. Hsiao, P.Kopp, and W.L. Wendland, *A Galerkin collocation method for some integral equations of the first kind*, Computing, **25**(1980), 89–130.

[35] G.C. Hsiao, P.Kopp, and W.L. Wendland, *Some applications of a Galerkin-collocation method for boundary integral equations of the first kind*, Math. Meth. in Appl. Sci., **6**(1984), 280–325.

[36] L.C. Hsu, *Selected Methods and Examples of Mathematical Analysis*, Shanwu Publisher, Shanghai, China, 1955.

[37] L.C. Hsu, *On a method for expanding multiple integrals in terms of integrals in lower dimensions,* Acta. Math. Acad. Sci. Hung., **14**(1963), 359–367.

[38] L.C. Hsu and Y.S. Chou, *An asymptotic formula for a type of singular oscillatory integrals,* Math. Comp., **37**(1981), 503–507.

[39] L.C. Hsu and T.X. He, *On the minimum estimation of the remainders in dimensionality lowering expansions with algebraic precision,* J. Math. (Wuhan), (1982), No. 2–3, 247–255.

[40] L.C. Hsu, R.H. Wang, and Y.S. Zhou, *On the boundary type quadrature formulas, the construction method and applications* Comp. Math. (China), 1978, No. 3, 54–75.

[41] L.C. Hsu and Y.J. Yang, *Dimensionality reducing expansion and boundary type quadrature formulas with algebraic precision*, Comp. Math. of Chinese Universities, **4**(1981), 361–369.

[42] L.C. Hsu and Y.S. Zhou, *Numerical integration in high dimensions,* Computational Methods Series. Science Press, Beijing, 1980.

[43] L.C. Hsu and Y.S. Zhou, *Two classes of boundary type cubature formulas with algebraic precision*, Calcolo, **23**(1986), 227–248.

[44] L.C. Hsu, Y.S. Zhou, and T.X. He, Topics on Numerical Integration in High Dimensions, Anhui Education Press, Hefei, 1985.

[45] P.M. Hummel and C.L. Seebeck, *A generalization of Taylor's theorem,* Amer. Math. Monthly, **56**(1949), 243–247.

[46] N. Obrechkoff, *Sur les Moyennes Arithmetiques de la Sĕrie de Taylor,* C.R. Acad. Sci. Paris, **210**(1940), 526–528.

[47] D.V. Ionescu, *Generalization of a quadrature formula of N. Obreschkoff for double integrals* (Romanian), Stud. Cerc. Mat., **17**(1965), 831–841.

[48] S. Jaffard and Ph. Laurençot, *Orthonormal Wavelets, Analysis of Operators, and Applications to Numerical Analysis*, Wavelet—A Tutorial in Theory and Applications, C.K. Chui (ed.), Academic Press, San Diego, 1992, 543–601.

[49] P. Keast and J.C. Diaz, *Fully symmetric integration formula for the surface of the sphere in S dimension,* SIAM J. Numer. Anal., **20**(1983), 406–419.

[50] M. Koshiba and M. Suzuki, *Application of boundary-element method to waveguide discontinuities*, IEEE-MTT, **34**(1986), 301–307.

[51] L.J. Kratz, *Replacing a double integral with a single integral,* J. Approx. Theory, **27**(1979), 379–390.

[52] M. Levin, *On a method of evaluating double integrals,* Tartu Riikl. Ül, Toimetised, **102**(1961), 338–341.

[53] M. Levin, *Extremal problems connected with a quadrature formula,* Eesti NSV Tead. Akad. Toimetised Füüs-Mat. Tehn. Seer., **12**(1963), 44–56.

[54] Y. Meyer, *Ondelettes et Opóerateurs*, Herman, Paris, 1990.

[55] H.N. Mhaskar, F.J. Narcowich, and J.D. Ward, *Quadrature formulas on spheres using scattered data*, preprint, 2000.

[56] S.G. Mikhlin, *Variational Methods in Mathematical Physics*, Pergamon, New York, 1964.

[57] N. Obreschkoff, *Neue Quadraturformeln,* Abhandl. d. preuss. Akad. d. Wiss., Math. Natur. wiss. K1., **4**(1940), 1–20.

[58] M.J.D. Powell, *Approximation Theory and Methods*, Cambridge University Press, London, 1981.

[59] F. Riesz and B. Nagy, *Functional Analysis*, Blackie, New York, 1956.

[60] M. Sadowsky, *A formula for the approximate computation of a triple integral,* Amer. Math. Monthly, **47**(1940), 539–543.

[61] X. Shen, *A quadrature formula based on sampling in Meyer wavelet subspaces*, 2000.

[62] X. Shi, *On the problem of asymptotic expansion for integrals of oscillatory functions*, Comp. Math. of Chinese Universities, **1**(1979), 120–122.

[63] X. Shi and Z. Lu, *Remainder estimates for integrals of oscillatory functions*, Comp. Math. (China), **2**(1980), No. 4, 379–382.

[64] M. Skopina, *Wavelet approximation of periodic functions,* J. Approx. Theory, **104**(2000), 302–329.

[65] D.D. Stancu, *Sur quelques formules generales de quadrature du type Gauss–Christoffel,* Mathematica (Cluj), **1**(1959), 167–182.

[66] D. D. Stancu and A.H. Stroud, *Quadrature formulas with simple Gaussian nodes and multiple fixed nodes,* Math. Comp., **17**(1963), 384–394.

[67] A.H. Stroud, *Approximate Calculation of Multiple Integrals*, Prentice-Hall, Englewood Cliffs, N.H., 1971.

[68] J.L. Walsh, *Interpolation and Approximation*, Amer. Math. Soc., Providence, R.I., 1956.

[69] G.G. Walter, *Wavelets and Other Orthogonal Systems with Applications*, CRC Press, Ann Arbor, 1994.

[70] X. Wang, *On the asymptotic expansion for integral* $\int f(x, \langle \lambda x \rangle)dx$ *and its remainder estimate,* J. of Math. Res. Expo., **3**(1983), 39–45.

[71] X. Wang and Y.S. Zhou, *Asymptotic expansion for oscillatory integrals with singular factor*, Ke Xue Tong Bao, **13**(1982), 829.

[72] W.L. Wendland, *Boundary element methods and their asymptotic convergence*, Theoretical Acoustics and Numerical Techniques, CISM Courses 277, P. Filippi (ed.), Springer-Verlag, New York, 1983, 135–216.

[73] W.L. Wendland, *On some mathematical aspects of boundary element methods for elliptic problems*, Mathematics of Finite Elements and Applications V, J. Whiteman (ed.), Academic Press, London, 1985, 193–227.

[74] W.L. Wendland, *On asymptotic error estimates for the combined boundary and finite element method*, Innovative Numerical Methods in Engineering, R.P. Shaw et al. (eds.), Springer-Verlag, Berlin, 1986, 55–70.

[75] B. Wendroff, *Theoretical Numerical Analysis*, Academic Press, New York, 1966.

[76] J.E. Wilkins, *An integration scheme of Maréchal,* Bull. Amer. Math. Soc., **55**(1949), 191–192.

[77] Y. Xu, *Orthogonal polynomials and cubature formulae on spheres and on balls*, SIAM J. Math. Anal., **29**(1998), 779–793.

[78] K. Yashiro and S. Ohkawa, *Boundary element method for electromagnetic scattering from cylinders*, IEEE-AP, **33**(1985), 383–389.

[79] Y.S. Zhou and T.X. He, *Higher dimensional Korkin theorem,* Acta of Hefei University of Technology, 1985, No. 2, 1–8.

[80] Y.S. Zhou and T.X. He, *A method for constructing a kind of boundary type quadrature formulas,* Acta Science Natural University of Jilin, 1983, No. 4, 40–46.

Index